PRACTICAL PROBLEMS in MATHEMATICS
for ELECTRONIC
TECHNICIANS, Sixth Edition

Delmar's *PRACTICAL PROBLEMS in MATHEMATICS* Series

- *Practical Problems in Mathematics for Automotive Technicians, 5e*
  Sformo & Moore
  Order # 0-8273-4622-0

- *Practical Problems in Mathematics for Carpenters, 7E*
  Harry C. Huth
  Order # 0-7668-2250-8

- *Practical Problems in Mathematics for Drafting and CAD, 3E*
  John C. Larkin
  Order # 0-8273-4624-7

- *Practical Problems in Mathematics for Electricians, 6E*
  Herman and Garrard
  Order # 0-7668-3897-8

- *Practical Problems in Mathematics for Electronic Technicians, 6E*
  Herman
  Order # 1-4018-2500-1

- *Practical Problems in Mathematics for Graphic Communications, 2e*
  Ervin A. Dennis
  Order # 0-8273-7946-3

- *Practical Problems in Mathematics for Health Occupations*
  Louise M. Simmers
  Order # 0-8273-6771-6

- *Practical Problems in Mathematics for Heating and Cooling Technicians, 3e*
  Russell B. DeVore
  Order # 0-8273-7948-X

- *Practical Problems in Mathematics for Industrial Technology*
  Donna Boatwright
  Order # 0-8273-6974-3

- *Practical Problems in Mathematics for Manufacturing, 4e*
  Dennis D. Davis
  Order # 0-8273-6710-4

- *Practical Problems in Mathematics for Masons, 2e*
  John E. Ball
  Order # 0-8273-1283-0

- *Practical Problems in Mathematics for Welders, 4e*
  Schell and Matlock
  Order # 0-8273-6706-6

PRACTICAL PROBLEMS in MATHEMATICS
for ELECTRONIC TECHNICIANS, Sixth Edition

Stephen L. Herman

DELMAR

THOMSON LEARNING ™

Australia   Canada   Mexico   Singapore   Spain   United Kingdom   United States

DELMAR

™

THOMSON LEARNING

Practical Problems in Mathematics for Electronic Technicians

Stephen L. Herman

Vice President, Technology and Trades SBU:
Alar Elken

Editorial Director:
Sandy Clark

Development:
Jennifer Luck

Marketing Director:
Maura Theriault

Channel Manager:
Fair Huntoon

Marketing Coordinator:
Brian McGrath

Production Director:
Mary Ellen Black

Production Manager:
Andrew Crouth

Production Editor:
Sharon Popson

Technology Project Manager:
David Porush

Technology Project Specialist:
Kevin Smith

Library of Congress Cataloging-in-Publication Data

Herman, Stephen L.
  Practical problems in mathematics for electronic technicians / Stephen L. Herman.— 6th ed.
        p. cm. — (Delmar's practical problems in mathematics series)
    ISBN 1-4018-2500-1
    1. Electronics—Mathematics.
  2. Electronics—Problems, exercises, etc. I. Title. II. Series.
    TK7862.H47 2003
    621.381'0151—dc22
                                2003015013

NOTICE TO THE READER

Publisher does not warrant or guarantee any of the products described herein or perform any independent analysis in connection with any of the production information contained herein. Publisher does not assume, and expressly disclaims, any obligation to obtain and include information other than that provided to it by the manufacturer.

The reader is expressly warned to consider and adopt all safety precautions that might be indicated by the activities herein and to avoid all potential hazards. By following the instructions contained herein, the reader willingly assumes all risks in connection with such instructions.

The publisher makes no representation or warranties of any kind, including but not limited to, the warranties of fitness for particular purpose or merchantability, nor are any such representations implied with respect to the material set forth herein, and the publisher takes no responsibility with respect to such material. The publisher shall not be liable for any special, consequential, or exemplary damages resulting, in whole or part, from the readers' use of, or reliance upon, this material.

Contents

Preface

In order to succeed in the electronics field, one must have a substantial background in mathematics. The sixth edition of *Practical Problems in Mathematics for Electronic Technicians* has been written to provide beginning students with these needed skills. The text starts with basic arithmetic and progresses through algebra and trigonometry. The explanations at the beginning of each unit have been expanded to provide the student with a better understanding of the concepts being presented. Most of the problems are word problems designed to encourage the student to use logical deduction to arrive at an answer. Many of these word problems are multistepped. Problems related to the electronics field are used throughout the text to help students understand electrical terms and practices.

In addition, an awareness of electronics symbols, basic circuits, component terminology, and calculator use is developed through the text. The answers to odd-numbered problems are provided in the back of the text along with a complete appendix.

WHAT IS NEW FOR THE SIXTH EDITION?

- Expanded coverage of scientific notation

- More examples of how to use a calculator to solve problems

- Additional information on RLC circuits

- A complete new unit on simultaneous equations that includes coverage of Kirchhoff's law with step-by-step illustrations on how to solve problems using simultaneous equations

To the Student

USING A CALCULATOR

Welcome to the study of mathematics for electronic technicians. You are entering an exciting, rapidly changing field. This workbook will help prepare you for a successful career. Before you start to work any of the problems, there are a few points you need to consider.

HELPFUL HINTS

1. Read carefully! Be sure that you understand the problem.

2. When applicable, (that is, for operations of addition and subtraction), be sure that all of the values in a problem have the same units of measure.

3. When a schematic is not given, it may be helpful to draw and label the circuit.

4. Look carefully at all subscripts. For example, current number one and current number two will be written I_1 and I_2 respectively. The voltage across resistor number four will be written E_{R_4}.

5. Be sure to use the appendix until a thorough knowledge of the numerous symbols and abbreviations has been achieved.

6. To help clarify the values in a problem, the specific symbols may be given. These will be written in parentheses following a term. For example, frequency (f), source voltage (E_s), power (P), and so on. In this text, all voltages will be designated with an E.

7. The Greek letter π (pi) represents a special number used in mathematics. To seven decimal places $\pi = 3.1415927$. In this text we will round this value to $\pi \cong 3.1416$.

8. Both English and metric units of measure are used in this workbook. Please refer to the appendix to answer any questions on units of measure.

9. Example problems with answers are provided for most units.

10. The answers to the odd-numbered Practical Problems are given at the end of the Workbook. Check these answers as you work through a unit.

Take your time, think through each problem, and good luck in your study of electronics.

SELECTING A CALCULATOR

Selecting an appropriate calculator to help in your study of electronics is extremely important. There are many types and models of calculators on the market. Which one is right for you? Which functions and features will be the most useful? The following guidelines will be helpful as you select a calculator:

- Size is an important factor to consider. The student wishing to carry a calculator to class may shop for one that will fit in a pocket. An engineer may want a larger desktop model. When your calculator is not in use, it should be stored in an appropriate case.

- Check the calculator's power requirements and power options. Some units may operate on nonrechargeable batteries for thousands of hours of operation. Rechargeable battery packs are also available. A power-saving feature is the automatic turn-off function which removes the display after a short period of nonuse. Similarly, after a few minutes of nonuse the calculator will automatically shut off.

- Most of the available electronic calculators are capable of storing numbers in a memory. In some cases, these numbers are cleared when the power is off. Other models have a constant memory that will retain these numbers when the power is off. This constant memory is extremely useful and should be seriously considered when selecting a calculator.

- Calculators are often programmable. While not used in this text, this mode of operation allows the user to design a specific program within the memory of the calculator. The owner's manual will contain directions for programming.

- Every calculator has the capability of performing specific operations or functions. These are executed by depressing the appropriate keys. Which functions should you look for? The table on the following page lists the functions you will use in your study of electronics. Also identified are the typical key symbols for each function. It should be pointed out that most scientific calculators are designed to perform these functions and many more.

- Be sure to try a few different models before making your final selection. Also, ask other students, faculty, and colleagues for their recommendations. Once you have purchased an electronic calculator, it is very important that you thoroughly study the owner's manual.

- The final considerations are cost and warranty. The total cost of the calculator and all accessories should be within your budget. In addition, read all available information to ensure that you understand the duration and coverage of the warranty. Finally, be sure that you understand your responsibilities in terms of sales receipts, original packaging, and repair procedures.

Take care of your calculator. As you work through this text the two of you will become inseparable!

Typical Key Symbols and Functions of a Scientific Calculator

Key(s)	Function(s)
$+$, $-$, $\times$, $\div$, $=$	Basic arithmetic
$+/-$	Change sign
π	Pi
$($, $)$	Parentheses
EE	Scientific notation
Eng	Engineering notation
STO , RCL , EXC	Memory or memories
x^2 , $\sqrt{x}$	Square and square root
y^x	y to the x power
$\frac{1}{x}$	Reciprocal
$\%$	Percent
log , ln x	Logarithm and natural logarithm
DRG	Degrees, radians, and graduations
sin , cos , tan	Trigonometric functions
P$\leftrightarrow$R	Polar and rectangular conversions
INV	Inverse function

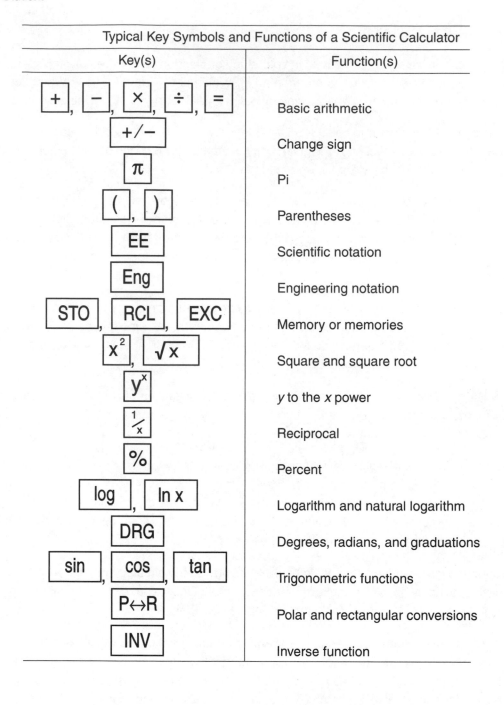

Whole Numbers

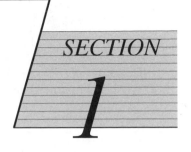

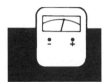

Unit 1   *ADDITION OF WHOLE NUMBERS*

BASIC PRINCIPLES OF ADDING WHOLE NUMBERS

Whole numbers are numerical units with no fractional parts. Addition is the process of finding the *sum* of two or more numbers. Whole numbers are added by placing them in a column with the numbers aligned on the right side of the column. The right column of numbers is added first. The last digit of the sum is written in the answer. The remaining digit is carried to the next column and added. This procedure is followed until all columns have been added.

Example: Find the sum of 18 + 22 + 276 + 31 + 422

		1		1
	18	18		18
	22	22		22
	276	276		276
	31	31		31
	+ 422	+ 422		+ 422
	9	69		769

PRACTICAL PROBLEMS

1. A company orders 425 resistors, 82 capacitors, 15 integrated circuits, and 48 transistors. What is the total number of parts ordered? _____

2. The military has a radio station that can broadcast 850 miles. With extra amplification it can reach another 325 miles. Find the total distance across which the station can transmit using the extra amplifcation. _____

3. A power supply contains five printed circuit boards. Separately they have 39, 62, 21, 43, and 58 components. What is the total number of components? _____

4. The values of resistors are added when the resistors are placed in series. Resistors of 470 ohms, 1,000 ohms, 39 ohms, 2,700 ohms, 680 ohms, and 10,000 ohms are placed in series. Find the total resistance. _____

Resistors are common in electronic equipment.

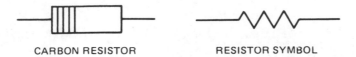

CARBON RESISTOR   RESISTOR SYMBOL

5. A technician connects wires with lengths of 4 meters, 7 meters, 28 meters, 16 meters, 30 meters, 83 meters, and 12 meters. Find the total length. _____

6. During the first half of the year a supply house shipped 4,800, 12,000, 8,700, 950, 15,000, and 15,000 fuses. Find the total number of fuses shipped. _____

Fuses protect electronic equipment.

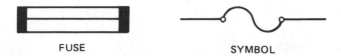

FUSE   SYMBOL

7. Kirchhoff's voltage law states that the sum of voltage drops around a closed loop equals source voltage. Find the source voltage (E_s) in this circuit. _____

    Note: $E_s = E_1 + E_2 + E_3 + E_4 + E_5$

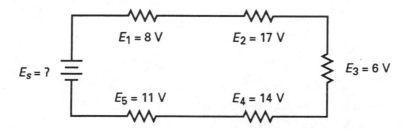

8. Kirchhoff's current law states the sum of currents into a point equals the current out of the point. What is the current out of the point in this diagram? _____

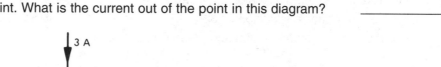

9. A live television broadcast from New York is sent 1,290 kilometers to Chicago. It is rebroadcast a distance of 803 kilometers to Kansas City and another 966 kilometers to Denver. From there it is sent 811 more kilometers to Salt Lake City and a final 1,151 kilometers to Los Angeles. What is the total distance the signal travels?   _____

10. A transformer may change electrical energy from one level to another. It is constructed by winding a number of turns of wire into a primary. Another set of turns is wound to form one or more secondaries. Find the total number of turns in the secondary of the transformer in this diagram.   _____

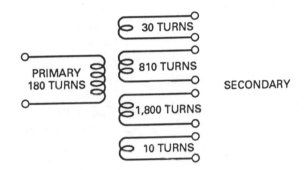

11. An electronic mixer can produce the sum of the two frequencies put into it. Frequencies of 1,145,000 hertz and 455,000 hertz are mixed. Find the output sum.   _____

12. The peak-to-peak voltage of an alternating current is found by adding the positive and negative peak voltages. Find the peak-to-peak voltage in this diagram.   _____

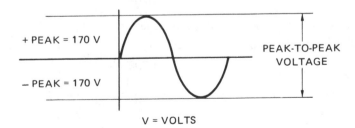

13. When capacitors are placed in parallel, the total capacitance is found by adding the individual capacitor values. Find the total capacitance (C_t) in picofarads (pF) in this circuit.

   _____

   Note: $C_t = C_1 + C_2 + C_3 + C_4 + C_5$

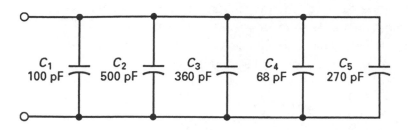

14. A delivery van travels 24 miles, 32 miles, 18 miles, 21 miles, and 16 miles during a week. What is the total distance traveled?

   _____

15. When inductors are connected in series the total inductance (L_t) is the sum of the individual inductors. Find the total inductance in millihenries (mH) in this circuit.

   _____

   Note: $L_t = L_1 + L_2 + L_3 + L_4$

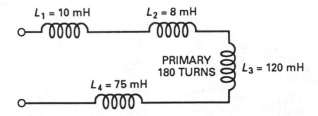

16. Some equipment is torn apart for the electronic parts mounted within it. The parts include 124 resistors, 87 capacitors, 22 diodes, 38 transistors, 12 integrated circuits, and 4 coils. What is the total number of components?

   _____

17. A manufacturer produces speakers for stereo systems. During the first half of the year, 8,700, 7,450, 8,464, 11,680, 9,750, and 9,453 speakers are produced. Find the total number of speakers produced.

   _____

Note: Use this drawing for problems 18 and 19.

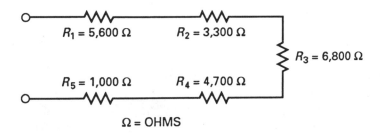

$R_1 = 5,600 \ \Omega$ $R_2 = 3,300 \ \Omega$

$R_3 = 6,800 \ \Omega$

$R_5 = 1,000 \ \Omega$ $R_4 = 4,700 \ \Omega$

$\Omega = $ OHMS

18. The diagram shows five resistors in series. The total resistance (R_t) is the sum of the individual resistances. Find the total resistance in ohms.   _____

   Note: $R_t = R_1 + R_2 + R_3 + R_4 + R_5$

19. Resistor R_2 in the diagram is changed to 10,000 ohms. Find the total resistance.   _____

20. The branch currents in a parallel circuit are added to form total current (I_t). Find I_t in this parallel circuit in milliamperes (mA).   _____

   Note: $I_t = I_1 + I_2 + I_3$

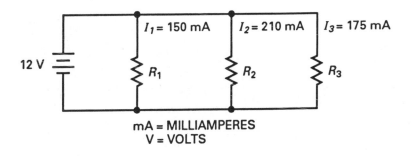

12 V

$I_1 = 150$ mA $I_2 = 210$ mA $I_3 = 175$ mA

R_1 R_2 R_3

mA = MILLIAMPERES
V = VOLTS

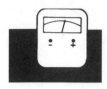

UNIT 2   SUBTRACTION OF WHOLE NUMBERS

BASIC PRINCIPLES OF SUBTRACTING WHOLE NUMBERS

To better understand the process of subtraction, refer to the number line that follows. The center of the number line has a value of zero. Numbers to the right of the zero have a positive value and numbers to the left of the zero have a negative value. Subtraction is basically the process of adding numbers with different signs.

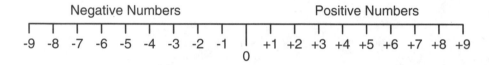

Example: Consider that 4 is to be subtracted from 6. This could actually be expressed as:

$$6 + (-4) = 2$$

To understand this concept, refer to the number line that follows. The positive 6 extends from the zero line six units to the right. Since the number 4 is negative, it extends four units to the left. The result is positive 2.

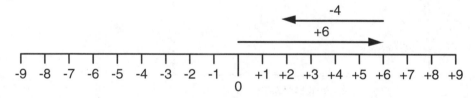

It is common practice to remove the negative number from the parentheses by changing the sign preceding the parentheses. Therefore, when the negative four is removed from the parentheses, the plus sign is changed to a minus sign as shown:

$$6 - 4 = 2$$

Using Signed Numbers

In common practice we think of subtraction as finding the difference between two numbers by subtracting the smaller number from the larger. It is also possible to add or subtract numbers with the same sign or different signs. To understand this concept, refer again to the first number line shown earlier. When a plus sign appears between two numbers, it means that they move in the same direction on the number line regardless of their signs. When a minus sign appears between two numbers, they move in the opposite direction on the number line.

Example: $-5 - (-3)$

As explained before, when the -3 is removed from the parentheses, the preceding sign changes. The new equation is:

$$-5 + 3 = -2$$

Refer to the number line that follows.

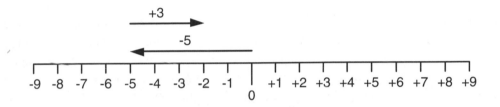

General Rule: When adding numbers that have different signs, subtract them and take the sign of the larger.

Now consider what happens if two negative numbers are added:

$$-5 + (-3)$$

becomes:

$$-5 - 3 = -8$$

Refer to the number line that follows.

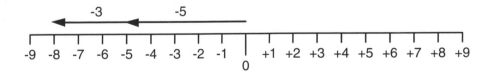

General Rule: When adding numbers with the same signs, add them and keep their sign.

Subtraction of Common Whole Numbers

Unless you are dealing with signed numbers, subtraction is accomplished by placing the smaller of the two numbers below the larger and finding the difference. It is important to keep the righthand column of numbers aligned.

Example: Subtract 524 from 896

$$
\begin{array}{r}
896 \\
- 524 \\
\hline
372
\end{array}
$$

Borrowing

In subtracting whole numbers, it is sometimes necessary to borrow from the number in the adjacent column. When you do this, the amount borrowed must be in increments of value of the column borrowed from. For example, starting from the right, the first column represents units or 1s, the second column represents 10s, the third column represents 100s, the fourth column represents 1,000s, etc. The number 9,876 could actually be rewritten as 1,000 nine times, 100 eight times, 10 seven times, and 1 six times.

1,000	100	10	1
1,000	100	10	1
1,000	100	10	1
1,000	100	10	1
1,000	100	10	1
1,000	100	10	1
1,000	100	10	
1,000	100		
1,000			
9,000	800	70	6

Now assume that the number 7,787 is to be subtracted from 9,876. Write the smaller number below the larger.

$$
\begin{array}{r}
9,876 \\
- 7,787 \\
\hline
\end{array}
$$

In this example, 7 cannot be subtracted from 6. Therefore, the 6 must borrow from the 7 in the column adjacent to it. Since the 7 is in the 10s column, 10 is borrowed, leaving 6 in that column. The borrowed 10 is added to the original 6, making 16 (10 + 6 = 16). Now 7 can be subtracted from 16, leaving a difference of 9.

$$986 \textbf{ (16)}$$
$$\underline{-\ 778\ (\ \textbf{7})}$$
$$9$$

In the next column, 80 must be subtracted from 60. Since this is not possible, 100 will be borrowed from the 800 in the adjacent column and added to the existing 60, making 160. (This now leaves 7 in the 100s column.) The difference will be 80.

$$97\ \textbf{(160)}\ 6$$
$$\underline{-\ 77\ (\ \textbf{80})\ 7}$$
$$8\quad 9$$

In the third column, 700 is subtracted from 700, leaving a difference of 0.

$$9,876$$
$$\underline{-\ 7,787}$$
$$089$$

In the fourth column, 7,000 is subtracted from 9,000, leaving a difference of 2,000.

$$9,876$$
$$\underline{-\ 7,787}$$
$$2,089$$

Another Borrowing Method

"Borrow 1" is another borrowing method used in the subtraction of whole numbers. The term is actually a misnomer because 1 can be borrowed only from the units column, but many people use this method and find it simpler to understand. Assume the number 58 is to be subtracted from the number 843. Place 58 below 843.

$$843$$
$$\underline{-\ 58}$$

Since 8 cannot be subtracted from 3, 1 is borrowed from the adjacent 4. The 3 now becomes 13, and the 4 becomes 3.

$$
\begin{array}{r}
8\ 3\ (13) \\
-\ 5\quad 8 \\
\hline
5
\end{array}
$$

The 5 must now be subtracted from the 3 in the second column. Since 5 cannot be subtracted from 3, 1 is borrowed from the adjacent 8, and the 3 becomes 13. The 8 now becomes a 7.

$$
\begin{array}{r}
7\ (13)\ 3 \\
-\quad 5\ 8 \\
\hline
7\ 8\ 5
\end{array}
$$

PRACTICAL PROBLEMS

1. An electronics supply store received 4,000 miniature light bulbs. During a sale 2,837 are sold. How many bulbs remain in stock?   _____

2. A power source supplies 48 amperes of current (I_t) to two cabinets of equipment. Cabinet A draws 23 amperes of current. How many amperes does Cabinet B draw?   _____

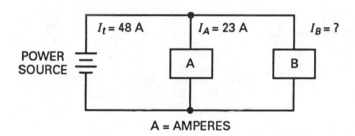

$I_t = 48$ A $I_A = 23$ A $I_B = ?$

POWER SOURCE   A   B

A = AMPERES

3. A television technician had 735 transistors in the shop. During a four-week period, 38, 27, 42, and 51 transistors are used. How many transistors are left in stock?   _____

4. A computer company charges $900 for a repair call. The cost includes $375 for new printed circuit boards, $190 to repair a terminal, and $84 in miscellaneous parts and overhead. Find the amount of profit.   _____

5. A potentiometer is a variable resistor. Its resistance can be changed from zero ohms of resistance to the maximum resistance of the potentiometer. A 10,000-ohm potentiometer is set to 3,500 ohms between lugs A and C. Find the ohms of resistance between lugs B and C if the maximum resistance of the potentiometer is to be achieved.

_____

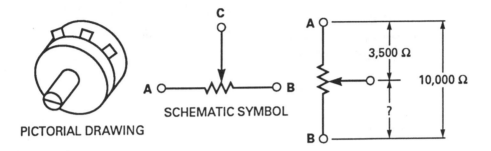

SCHEMATIC SYMBOL

PICTORIAL DRAWING

6. An electronic mixer can produce the difference between two input frequencies. The input signals are 1,200,000 hertz and 745,000 hertz. Find the difference frequency.

_____

7. A television manufacturer builds 347,000 printed circuit boards. During a six-month period, 38,800, 45,627, 53,752, 43,650, 58,733, and 34,975 circuit boards are used to construct television sets. How many boards are left in stock?

_____

8. The values of resistors in series are added to form total resistance. The total resistance (R_t) shown in this schematic is 5,470 ohms. Find the value of R_4.

_____

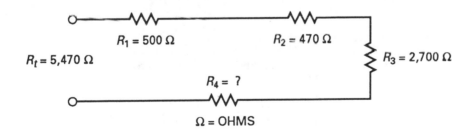

Ω = OHMS

9. A television antenna must be 35 meters high. A 10-meter section and a 12-meter section are used. How long should the third section be to complete the antenna?

_____

10. A 500-foot roll of wire is used to install an intercom in a new house. To wire the basement, the first floor, and the second floor, 37 feet, 26 feet, and 74 feet of wire are used. How many feet of wire are left on the spool?

_____

11. To pay for VCR servicing, Emily wrote a check for $37. She also wrote a check for $15 to purchase 3 tapes. Her checkbook balance was $185 before writing the checks. How much money remains?

_____

12. The peak-to-peak voltage of a square wave is 35 volts. The upper portion of the wave is 14 volts. Find the value of the lower portion of the wave.

_____

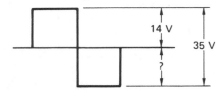

13. A cable television company has 100,000 feet of wire to connect the houses of new customers. During a four-week period, 7,800 feet, 8,400 feet, 10,750 feet, and 9,379 feet of wire are installed. How many feet of wire remain?

_____

14. A computer assembler receives a payroll check with a gross pay of $865. After deducting $161 for federal tax, $15 for state tax, and $73 for retirement, what is the net pay?

_____

15. A technician is given a $6,000 budget to purchase some test equipment. Refer to the following chart and determine how much money remains after the equipment is purchased.

_____

ITEM	QUANTITY	UNIT COST (in dollars)	TOTAL COST (in dollars)
Meter	3	150	450
Oscilloscope	2	1,200	2,400
Signal generator	2	410	820
Transistor tester	1	350	350
Power supply	4	280	1,120
Tool set	1	185	185

16. A repair technician has a supply of 160 integrated circuits. During four major repair jobs, 38, 27, 19, and 47 circuits are used. How many circuits are unused?

_____

17. Voltage drops in a series circuit are added to give total source voltage. The source voltage (E_S) is 48 volts. Find the voltage across R_2.

_____

   Note: $E_{R_2} = E_S - E_{R_1} - E_{R_3} - E_{R_4}$

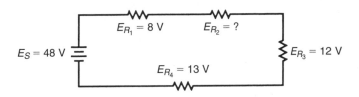

18. A delivery truck has 486 miles to travel. The driver goes 112 miles and stops for lunch. After another 63 miles, the driver stops for gas. How many miles are left to drive?

_____

19. A power supply is designed to produce 400 volts. It is set to 287 volts. How many more volts can it produce?

_____

20. The branch currents in a parallel circuit are added to equal total source current (I_t). Total current in the circuit is 175 milliamperes. Find the current in branch three (I_3).

_____

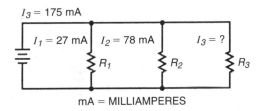

Unit 3 MULTIPLICATION OF WHOLE NUMBERS

BASIC PRINCIPLES OF MULTIPLYING WHOLE NUMBERS

Multiplication is actually a method of addition used when like numbers are added. For example, if four 5s are added, the answer will be 20. If the number 5 is multiplied by 4, the answer (known as the *product*) is equal to 20. Therefore, 5 × 4 is the same as adding four 5s.

$$\begin{array}{r} 5 \\ 5 \\ 5 \\ +\ 5 \\ \hline 20 \end{array} \qquad \begin{array}{r} 5 \\ \times\ 4 \\ \hline 20 \end{array}$$

To multiply larger numbers, first write the number to be multiplied; then write underneath it the number of times it is to be multiplied. In the following example, the number 247 is to be multiplied by 32. Write the numbers keeping the units column aligned.

Example: 247 × 32

$$
\begin{array}{r} 1 \\ 247 \\ \times\ 32 \\ \hline 4 \end{array}
\qquad
\begin{array}{r} 1 \\ 247 \\ \times\ 32 \\ \hline 94 \end{array}
\qquad
\begin{array}{r} 1 \\ 247 \\ \times\ 32 \\ \hline 494 \end{array}
\qquad
\begin{array}{r} 2 \\ 1 \\ 247 \\ \times\ 32 \\ \hline 494 \\ 1 \end{array}
\qquad
\begin{array}{r} 12 \\ 1 \\ 247 \\ \times\ 32 \\ \hline 494 \\ 41 \end{array}
\qquad
\begin{array}{r} 12 \\ 1 \\ 247 \\ \times\ 32 \\ \hline 494 \\ 7\ 41 \\ \hline 7,904 \end{array}
$$

Multiply the units (2 × 7 = 14). Place the 4 below the 2 and carry the 1 to the next column. Then multiply (2 × 4 = 8) and add the 1 (8 + 1 = 9). Place the 9 beside the 4. Multiply (2 × 2 = 4). Place the 4 beside the 9. Next, multiply each digit of 247 by the 3 in 32. The answers will be brought down in the same manner except that one space is skipped. Multiply (3 × 7 = 21). Place the 1 below the 9 and carry the 2 to the next column. Multiply (3 × 4 = 12). Add the 2 from the first column (12 + 2 = 14). Place the 4 beside the 1 and carry the 1 to the next column. Multiply (3 × 2 = 6) and add 1 (6 + 1 = 7). Place the 7 beside the 4. The final step is to add the two sets of products together to obtain the total.

PRACTICAL PROBLEMS

1. A technician uses 12 pieces of wire. Each piece is 7 inches long. How many inches of wire are used?   _____

2. A parts delivery van travels at 55 miles per hour. How far has it gone after 4 hours?   _____

3. The current gain of a common emitter transistor amplifier is constant. To find the collector current (I_c), multiply the current gain (β) times the base current (I_b). The base current is 75 microamperes (μA), and the gain is 65. Find the collector current in microamperes.   _____

    Note: $I_c = \beta \times I_b$

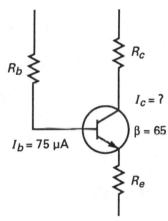

4. Find the collector current (I_c) when the current gain (β) is 40 and the base current (I_b) is 25 microamperes.   _____

5. Three workers spent 6 hours installing an antenna. They are each paid $15 per hour.

    a. Find the amount each worker earns.   a. _____

    b. Find the total earnings.   b. _____

6. A certain type of wire has a resistance of 2 ohms per 1,000 feet. Find the total resistance of 34,000 feet of this wire.   _____

7. Ohm's law states that voltage (E) equals current (I) times resistance (R). What is the voltage drop in volts across a 12-ohm resistor if it is drawing 2 amperes of current?

   Note: $E = I \times R$

   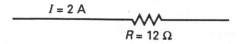

8. Find the voltage drop (E) in volts across a 5-ohm resistor (R) if it is drawing 25 amperes of current.

9. A roll of wire weighs 20 pounds. This type of wire contains 348 feet of wire per pound. Find the total length of wire on the roll.

10. Power, the rate of doing work, is measured in watts. Power (P) is equal to the voltage (E) across a device times current (I) through the device. Find the power in watts consumed by a load that draws 3 amperes of current with an 18-volt drop across it.

    Note: $P = I \times E$

    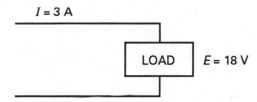

11. Find the power in watts consumed by a load with a voltage drop (E) of 25 volts and a current (I) of 5 amperes passing through it.

12. A 12-volt car battery supplies 250 amperes of current to the starter motor. What is the power in watts consumed by the motor?

13. Each vertical centimeter on an oscilloscope screen is equal to a number of volts. A sine wave is 4 centimeters from peak-to-peak. Each centimeter is equal to 25 millivolts. Find the total peak-to-peak voltage. _____

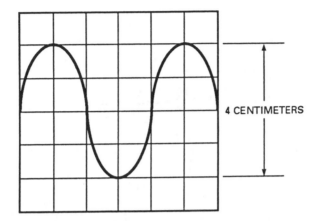

4 CENTIMETERS

14. A 10-to-1 probe is attached to a voltmeter to read larger than normal voltages. The actual voltage will be 10 times that shown on the meter. The meter reads 82 volts. What is the actual voltage being measured? _____

15. An oscillator generates an alternating current signal at a frequency determined by its components. The oscillator's output is 5,500 cycles per second (C/s). How many cycles does it produce in 35 seconds? _____

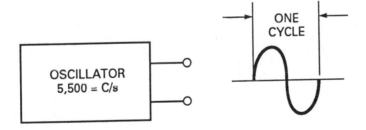

OSCILLATOR
5,500 = C/s

ONE
CYCLE

16. An electronics company produces 34 complete units each day. How many units are produced in 30 days? _____

17. The output voltage of an amplifier is found by multiplying the input voltage by the amplifier's voltage gain. The input is 20 millivolts and the gain is 25. Find the output voltage in millivolts. _____

18.   A 68-ohm resistor (R) draws 3 amperes of current (I). Find the voltage drop (E) across the resistor.

   Note: $E = I \times R$

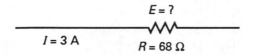

19.   A radio station plays 16 songs each hour, 24 hours each day. How many songs are played in 7 days?

20.   A lamp is connected to 120 volts (E). The bulb draws 2 amperes of current (I). Find the power (P) in watts consumed by the bulb.

   Note: $P = I \times E$

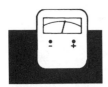

Unit 4   DIVISION OF WHOLE NUMBERS

BASIC PRINCIPLES OF DIVIDING WHOLE NUMBERS

In Unit 3 it was shown that multiplication is actually the process of adding a number to itself a certain number of times. Division is just the opposite or inverse. Division is actually the process of subtracting a smaller number from a larger number a certain number of times. The larger number, the number to be divided, is referred to as the *dividend*. The number used to indicate the number of times the dividend is to be divided is called the *divisor*. The answer is known as the *quotient*.

To begin the process of division, the dividend is placed inside the division bracket and the divisor is placed to the left of the dividend. The quotient is placed above the dividend.

$$\text{Divisor)} \overline{\begin{array}{c} \text{Quotient} \\ \text{Dividend} \end{array}}$$

Example: Divide 1225 by 18 (1225 ÷ 18)

To begin the process of division, place the dividend under the division bracket and the divisor to the left of the bracket.

```
                    4    6        6    68              68 r 1
  18) 1225       18) 1225      18) 1225            18) 1225
                    108           108                 108
                    145           145                 145
                                  144                 144
                                    1                   1
```

The divisor cannot be divided into a number smaller than itself. Therefore, the number 18 cannot be divided into 1 or 12, but it can be divided into 122. 18 will divide into 122 a total of 6 times. Place 6 above the division bracket directly above the second 2 in the number 122. Now multiply 6 times the divisor (6 × 8 = 48). Place the 8 beneath the second 2 in 122 and carry the 4. (6 × 1 = 6; 6 + 4 = 10.) Place the 10 beside the 8 and the number is 108. Subtract 108 from 122. This produces an answer of 14. Bring down the next number of the dividend (5) and place it beside the 14. The number is now 145. Divide 18 into 145. 18 will divide into 145 a total of 8 times. Place the 8 in the quotient beside the 6 and multiply the divisor by 8 (8 × 8 = 64). Place the 4 under the 5 in 145 and carry the 6. (8 × 1 = 8; 8 + 6 = 14.) Place the 14 beside the previous 4 to make the number 144. Subtract 144 from 145. This produces an answer of 1. Since the dividend has no other integers to bring down, the quotient is 68 with a remainder of 1 (68 r 1).

PRACTICAL PROBLEMS

1. A supply truck travels 468 miles in 9 hours. At what speed in miles per hour
    is the truck driven?

    _____

2. Current gain (β) of a transistor amplifier equals the output current (I_c) divided
    by the input current (I_b). In this circuit, I_c is 25 microamperes. Find β in the
    circuit shown.

    Note: $\beta = I_c \div I_b$

    _____

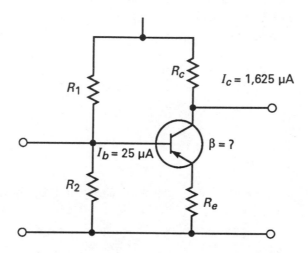

3. A technician receives a gross payroll check of $1,056. This check represents
    11 working days, at 8 hours per day.

    a. Find the amount of pay per day.   a. _____

    b. Find the amount of pay per hour.   b. _____

4. The wires on the back of a picture tube are 20 inches long. How many can
    be cut from a 500-foot roll?

    Note: 1 foot = 12 inches

    _____

5. An electronics warehouse receives 446 stereo plugs. The plugs are then packed in sets of 3 selling at $2 per set.

    a. Find the number of full sets packed.   a. _____

    b. Find the total amount of money all the sets sell for.   b. _____

    c. Find the number of plugs remaining.   c. _____

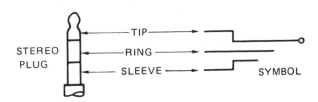

6. Ohm's law states resistance (R) in ohms is equal to voltage (E) divided by current (I). Twenty-four volts are applied to a circuit in which 3 amperes of current exist. Find the resistance in ohms.   _____

    Note: $R = E \div I$

7. Ruth is a ham operator. She can broadcast 5 times the distance that Jerry can broadcast. Ruth reaches a distance of 2,760 miles. How many miles does Jerry broadcast?   _____

8. A company manufactures 6,500 radios. The total income is $247,000. Find the price of each radio.   _____

9. Bryan is a disc jockey who is on the air 6 hours each day, 5 days each week. He logs his hours of broadcast time. His record shows 4,320 hours.

    a. Find the number of days worked.   a. _____

    b. Find the number of weeks worked.   b. _____

10. When batteries are connected in series, the total voltage is the sum of the individual voltages. How many 12-volt batteries are needed to obtain 276 volts?   _____

11. To assemble a stereo kit requires 42 hours. Working 3 hours each day, how many days does it take a person to assemble the stereo?   _____

12. A warehouse receives 38,000 transistor sockets. They are divided equally into 16 storage bins. How many are in each bin?   _____

13. A tank used to etch printed circuit boards is filled with a solution. It takes 52 quarts to fill the tank. Find the number of gallons needed to fill the tank.

   Note: 4 quarts = 1 gallon

   _____

14. The voltage gain of an amplifier is the output voltage divided by the input voltage. The input is 35 millivolts and the output is 1,470 millivolts. Find the gain.

   _____

15. Total current splits into branches in a parallel circuit. Total current (I_t) in the diagram is 900 milliamperes. The current splits equally into the branches. Find the current in one branch.

   Note: $I_1 = I_2 = I_3 = I_4 = I_5 = I_6$

   _____

I_t = 900 mA

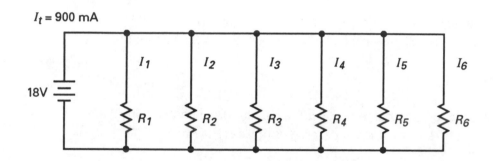

Note: The following symbols and equations are used in problems 16–20.
   E = voltage in volts R = resistance in ohms
   I = current in amperes P = power in watts

   $I = E \div R$ $R = E \div I$ $I = P \div E$ $E = P \div I$

16. A resistor is connected to 15 volts. If 3 amperes of current exist, find the resistance of the circuit in ohms.

   _____

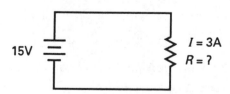

17.   A load is consuming 100 watts of power. Find the applied voltage if the load
current is 1 ampere.   _____

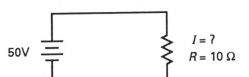

18.   A 10-ohm resistor is connected to 50 volts. Find the current in amperes.   _____

19.   A 2-ohm resistor is consuming 72 watts of power. It is connected to 12 volts.
Find the current.   _____

20.   The load in this circuit has an 8-volt drop across it. The current is 2 amperes.
Find the load resistance in ohms.   _____

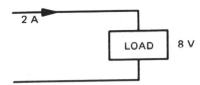

Unit 5 COMBINED OPERATIONS WITH WHOLE NUMBERS

The four basic functions—addition, subtraction, multiplication, and division—are used to solve many problems in electronics. Even when problems involve more advanced forms of mathematics, some portion includes these four operations, either alone or in conjunction with advanced mathematical functions.

PRACTICAL PROBLEMS

1. An electronics firm produces integrated circuits. During the first half of the year, 12,820, 10,500, 14,795, 12,700, 13,242, and 16,908 circuits are shipped. Find the total number of circuits shipped. _____

INTEGRATED CIRCUIT

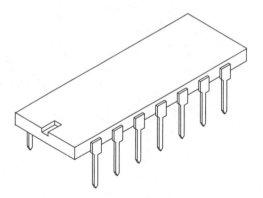

2. During a 5-week sale, a supply house sells an average of 82 speakers per week. There were 524 speakers in stock. How many are remaining after the sale? _____

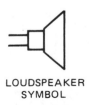

LOUDSPEAKER
SYMBOL

3. A television technician's October income includes 22 working days. Nine of these days are 8 hours long. The remaining days are 7 hours long. The technician is paid $17 per hour. Find the total income before deductions. _____

4. The values of resistors connected in series are added to find total resistance. A circuit requires 175,600 ohms of resistance. Resistors of 68,000 ohms, 22,000 ohms, 33,000 ohms, and 5,600 ohms are in stock. What additional value of resistance is needed? _____

Note: The following symbols and equations are used in problems 5–10.

E = voltage in volts $\qquad$ R = resistance in ohms

I = current in amperes $\qquad$ P = power in watts

$$E = I \times R \qquad\qquad P = I \times E$$
$$I = E \div R \qquad\qquad I = P \div E$$
$$R = E \div I \qquad\qquad E = P \div I$$

5. A resistor receives 2 amperes of current. It has a 30-volt drop. Find the resistance in ohms. _____

6. The heating element in a piece of equipment has a resistance of 16 ohms. At a voltage of 48 volts, what current is passing through the element? _____

7. The load in this circuit is consuming 24 watts of power. Find the amount of applied voltage. _____

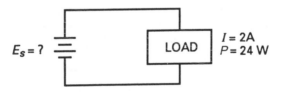

8. A motor connected to 120 volts draws a current of 4 amperes. What is the motor's power consumption in watts? _____

9. A 22-ohm resistor has 1 ampere of current passing through it. Find the voltage drop across the resistor. _____

10. A load is consuming 100 watts of power. The source voltage is 20 volts.

   a. Find the amount of current in amperes.　　　　a. _____

   b. Find the load resistance in ohms.　　　　b. _____

11. A delivery driver travels 55 miles per hour for 3 hours. Due to construction, the next 2 hours are driven at 45 miles per hour. The driver has a total distance of 290 miles to travel. How many miles remain to travel?   _____

12. A container of 12 boxes of fuses costs $84. What is the cost of a single box?   _____

13. A voltmeter has a 5,000-ohm-per-volt rating. This means that on the 3-volt range, the meter has $3 \times 5,000$, or 15,000, ohms of resistance. What is the resistance on the 25-volt range?   _____

14. A power generator produces a voltage at a frequency of 60 cycles per second. This is a frequency of 60 hertz. How many cycles are generated per hour?   _____

15. The forward-biased resistance of a diode is 15 ohms. The reverse-biased resistance is 700 times the forward resistance. Find the reverse resistance.   _____

DIODE   SCHEMATIC SYMBOL

16. A quality control technician determines that 1 in 24 VCRs is defective. If there are 864 VCRs to inspect, how many unsatisfactory VCRs are expected?   _____

17. A project requires wires of different lengths. Use the following table to determine the total amount of wire needed.   _____

NUMBER OF WIRES	LENGTH PER WIRE
4	10 IN
7	14 IN
2	27 IN
6	8 IN

18. Sound moves through the air at about 1,100 feet per second. How far does a sound wave travel in 8 seconds?   _____

19. A television set is used an average of 5 times each day. How many times is it used in 3 years? _____

    Note: There are 365 days in a year.

20. An antenna on a broadcast tower sends a signal to an area sweeping 235° around the station. How much of the area in degrees is not covered? _____

    Note: There are 360° in a circle.

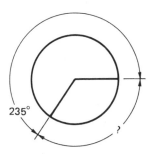

21. Light travels at a speed of 186,000 miles per second. It takes light approximately 8 minutes to reach the earth from the sun. What is the approximate distance from the earth to the sun? _____

22. Radar operates by broadcasting a high-frequency radio signal that travels at the speed of light. When the signal strikes an object, it is reflected and bounces back to its source. The distance of the object can be determined by the amount of time that elapses between the instant that the signal leaves the antenna and returns. Assume that the amount of time that elapses between a radar signal being sent and bouncing back is 0.001613 second. One minute later, the elapsed time between the antenna and the object is 0.001555 second.

    a. What is the distance of the object the first time the signal is sent?   a. _____

    b. What is the distance the second time?   b. _____

    c. What is the speed of the object, with respect to the radar operator, expressed in miles per hour?   c. _____

Common Fractions

Unit 6   ADDITION OF
COMMON FRACTIONS

BASIC PRINCIPLES OF ADDING COMMON FRACTIONS

Measurement often must be more precise than can be done with whole numbers. One method of indicating quantities that are smaller than a whole is with common fractions. There are two parts to a common fraction, the numerator (the number above the line) and the denominator (the number below the line.)

The denominator indicates the number of equal parts the whole is divided into. The inch, for example, is often divided into 16 equal parts. The numerator indicates the number of parts used. If a measurement is $\frac{5}{16}$ inch, it indicates that an inch has been divided into 16 equal parts and the length corresponds to 5 of these 16 parts.

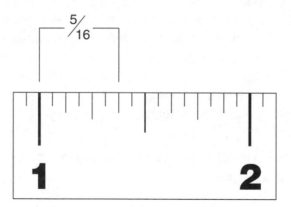

Adding Fractions

When fractions are added, their denominators must all be the same. When all the denominators are the same, only the numerators are added.

Example: $\frac{1}{16} + \frac{3}{16} + \frac{5}{16}$

$$
\begin{array}{r}
\frac{1}{16} \\[4pt]
\frac{3}{16} \\[4pt]
+\frac{5}{16} \\[4pt]
\hline
\frac{9}{16}
\end{array}
$$

Finding a Common Denominator

If all the denominators are not the same, it is necessary to find a *common denominator.* A common denominator can be found by finding some number that all the denominators of the individual fractions will divide. To say a number *divides* another means no remainder is left in the division. In the following example, a common denominator could be 24, since the denominator of each fraction will divide 24. This is not to say that 24 is the only common denominator. Another common denominator is 48, for example. However, 24 is the lowest common denominator (LCD), and use of the LCD is preferred when adding fractions.

Example: $\frac{5}{12} + \frac{1}{4} + \frac{1}{6} + \frac{3}{24}$

$$
\begin{array}{ccccc}
\frac{5}{12} & \times & \frac{2}{2} & = & \frac{10}{24} \\[6pt]
\frac{1}{4} & \times & \frac{6}{6} & = & \frac{6}{24} \\[6pt]
\frac{1}{6} & \times & \frac{4}{4} & = & \frac{4}{24} \\[6pt]
+\frac{3}{24} & \times & \frac{1}{1} & = & +\frac{3}{24} \\[6pt]
& & & & \hline \\[-6pt]
& & & & \frac{23}{24}
\end{array}
$$

Addition of these fractions is accomplished by changing each into an *equivalent fraction* with a denominator of 24. This is done by dividing the denominator of the fraction to be changed into the common denominator, and then multiplying the numerator by the answer. For example, $\frac{5}{12}$ is changed to $\frac{10}{24}$ because $24 \div 12 = 2$ and $2 \times 5 = 10$. The fractions $\frac{5}{12}$ and $\frac{10}{24}$ are equal in value. To change $\frac{1}{4}$ into an equivalent fraction in 24ths, note that $24 \div 4 = 6$ and $6 \times 1 = 6$. Therefore, the fraction $\frac{1}{4}$ has the same value as $\frac{6}{24}$. Once each fraction has been changed into an equivalent fraction in 24ths, the numerators are added together.

Finding a common denominator for the fractions in the preceding example was relatively simple because it was obvious that all denominators divided 24. There may be occasions, however, when a common denominator is not apparent. Assume that it is necessary to add $\frac{8}{17}$, $\frac{1}{5}$, and $\frac{3}{11}$. A number that can be divided by 17, 5, and 11 is not immediately apparent. A common denominator can always be found, however, by multiplying all the denominators together. This may not reveal the lowest common denominator, but it will produce a common denominator.

Example: $\frac{8}{17} + \frac{1}{5} + \frac{3}{11}$; LCD = $17 \times 5 \times 11 = 935$. Hence,

$$\frac{8}{17} = \frac{440}{935}$$

$$\frac{1}{5} = \frac{187}{935}$$

$$+ \frac{3}{11} = \frac{+255}{935}$$

$$\frac{882}{935}$$

Since $935 \div 17 = 55$ and $55 \times 8 = 440$; $935 \div 5 = 187$ and $187 \times 1 = 187$; and $935 \div 11 = 85$ and $85 \times 3 = 255$.

Reducing Fractions to Lowest Terms

It is common practice to reduce a fraction to its lowest terms. This is done by dividing both the numerator and denominator by the same number.

Example: Express $\frac{18}{48}$, in lowest terms.

$$\frac{18 \div 6 = 3}{48 \div 6 = 8}$$

The fractions $\frac{3}{8}$ and $\frac{18}{48}$ are equal in value. Some fractions, such as $\frac{23}{24}$, cannot be reduced because there is no number except 1 that will divide into both 23 and 24. The fraction $\frac{23}{24}$ is in lowest terms.

When adding fractions, it is not uncommon for the answer to include a fraction with a numerator greater than the denominator, such as $\frac{76}{24}$. This is called an *improper* fraction, which should be reduced to lowest terms. The first step is to divide the numerator by the denominator. This will produce a whole number of some value, and possibly a *remainder*.

$$\overset{\textstyle 3\ r\ 4}{24\overline{)\ 76}}$$

The second step is to rewrite the fraction as a *mixed* number. The fraction can now be rewritten as $3\frac{4}{24} = 3\frac{1}{6}$.

Note that the remainder in the division becomes the numerator in the fractional part of the mixed number.

USING A CALCULATOR WITH COMMON FRACTIONS

Many scientific type calculators have the ability to add, subtract, multiply, and divide common fractions. Calculators that have this ability generally have a key similar to the one shown here.

ab/c

Example: Add ¼ + ⅓ + ⅙ + ⅜

To add these fractions, press the following calculator keys:

1 ab/c 4 + 1 ab/c 3 + 1 ab/c 6 + 3 ab/c 8 =

The calculator display should indicate the following:

1⎵1 / 8

The answer is 1⅛.

Mixed numbers can also be computed with the calculator.

Example: Add 3⅕ + 4⅝ by pressing the following keys:

3 ab/c 1 ab/c 5 + 4 ab/c 5 ab/c 8 =

The calculator display should indicate the following:

7⎵33 / 40

The answer is 7³³/₄₀.

PRACTICAL PROBLEMS

1. A technician works 3¼ hours on Monday repairing a stereo. The technician works an additional 4½ hours Tuesday morning and 3½ hours Tuesday afternoon. Find the total hours worked.

    _____

2. The relay in the following diagram is placed in a socket on a circuit board. The board and relay must slide into a cabinet. Find the total vertical space required for the relay, socket, and board.

    _____

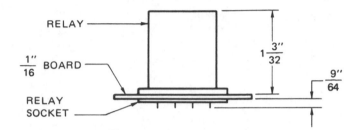

3. During a five-day workweek a technician works 9¼ hours, 8 hours, 8¾ hours, 9½ hours, and 7¼ hours. Find the total hours worked.

    _____

4. A television antenna has four sections. The sections are 6¾ feet, 12½ feet, 10¾ feet, and 14¼ feet long. When all sections are joined, how high is the antenna?

    _____

5. Voltages in series add to form a total voltage. Voltages of 38⅛ volts, 27⅝ volts, and 41½ volts are in series. Find the total voltage.

    _____

6. This metal plate is used to mount electronic components. How wide is the plate?

    _____

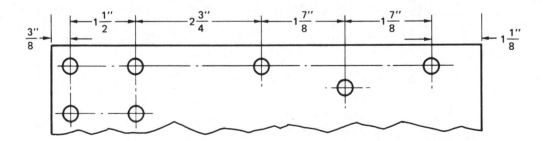

7. A solution is used to etch printed circuit boards. It contains 1⅔ gallons of water and 4½ gallons of a concentrate. How many gallons are in the mixture? _____

Note: Use this diagram for problems 8–10.

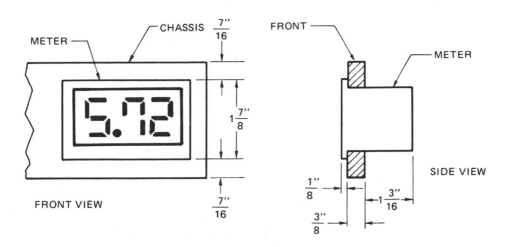

8. Refer to the meter and chassis drawing and find the overall depth of the meter. _____

9. Find the height of the chassis front. _____

10. A meter with the same front dimensions is installed. It extends 1¹³⁄₃₂ inches beyond the chassis instead of 1³⁄₁₆ inches. Find the overall meter depth. _____

11. While building a piece of test equipment, a technician cuts off six pieces of solder. The pieces are 4¾ inches, 6⅞ inches, 5⅛ inches, 9⁷⁄₁₆ inches, 7²³⁄₃₂ inches, and 7¹¹⁄₁₆ inches long. Find the total length of solder. _____

12. Two relay contacts are spaced ³⁄₃₂ inch apart. They are placed an additional ³⁄₆₄ inch apart. Find the total distance between the contacts. _____

Note: Use this diagram for problems 13 and 14.

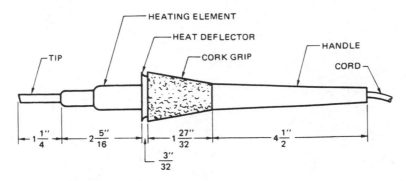

13.   Find the total length of the soldering iron pictured in this diagram.   _____

14.   The heating element is increased to 23⁷⁄₆₄ inches. Find the length of the soldering iron if all other dimensions are the same.   _____

15.   A spacer is sometimes placed under a transistor to prevent it from touching the printed circuit board. How far above the top surface of the board is the top of the transistor?   _____

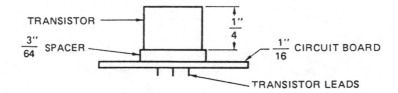

16.   A technician connects some wires together. The wires are 3¾ inches, 5⅞ inches, 2¹⁄₁₆ inches, 3¹⁷⁄₃₂ inches, and 6½ inches long. What is the combined length?   _____

17.   The perimeter of an object is the distance around the outside edges. Find the perimeter of this plate.   _____

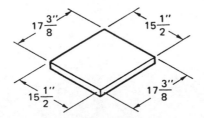

Note: Use this diagram for problems 18–20.

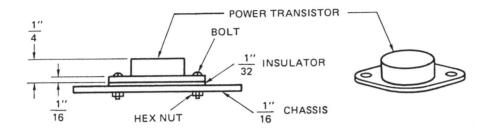

18.   What is the distance from the top of the transistor to the bottom of the chassis plate?   _____

19.   A hex nut is ³⁄₃₂ inch high. The bolt must extend ¹⁄₁₆ inch beyond the nut. Find the distance from under the head to the end of the bolt.   _____

20.   The same transistor and insulator are to be mounted on a ⁵⁄₃₂-inch chassis. Find the distance from the top of the transistor to the bottom of the plate.   _____

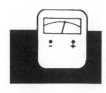

Unit 7   SUBTRACTION OF COMMON FRACTIONS

BASIC PRINCIPLES OF SUBTRACTING COMMON FRACTIONS

Subtraction of common fractions is similar to addition of fractions. In both operations, it is first necessary to have common denominators. Once both fractions have a common denominator, the numerator of the smaller fraction can be subtracted from the numerator of the larger.

Example: $\frac{5}{8} - \frac{7}{16}$

$$
\begin{array}{ccccc}
\dfrac{5}{8} & \times & \dfrac{2}{2} & = & \dfrac{10}{16} \\[2ex]
-\ \dfrac{7}{16} & \times & \dfrac{1}{1} & = & \dfrac{7}{16} \\[2ex]
& & & & \dfrac{3}{16}
\end{array}
$$

In this example, $\frac{7}{16}$ is subtracted from $\frac{5}{8}$. The lowest common denominator for these two fractions is 16. The fraction $\frac{5}{8}$ is changed to $\frac{10}{16}$: ($16 \div 8 = 2$; $2 \times 5 = 10$). The fraction $\frac{7}{16}$ is not changed. The numerator of the second fraction is then subtracted from that of the first ($10 - 7 = 3$). The answer is $\frac{3}{16}$.

PRACTICAL PROBLEMS

1. The lead-in wire on a television antenna is 37½ feet long. If 6¾ feet are cut off, how many feet are remaining?

   _____

2. A piece of metal 12¼ inches long is cut to make a chassis. The required length is 8¹⁵⁄₁₆ inches. How many inches should be cut off?

   _____

3. Fuses will stop the current in a circuit if the current exceeds the value of the fuse. The current in the circuit is 1⅞ amperes. How many more amperes of current will cause the fuse to blow?

   _____

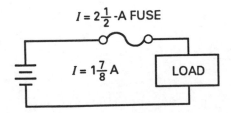

$I = 2\frac{1}{2}$ -A FUSE

$I = 1\frac{7}{8}$ A

LOAD

4. During a five-day workweek, a technician works 43¾ hours. The hours worked on the first four days are 7¼ hours, 8½ hours, 8¾ hours, and 8¾ hours. Find the number of hours worked on the fifth day.   _____

Note: Use this diagram for problems 5–8.

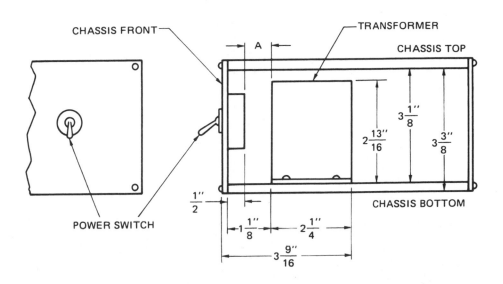

5. How much clearance is there between the transformer top and the chassis top?   _____

6. What is the distance from the transformer to the back of the switch, dimension A?   _____

7. Find the thickness of the front plate of the chassis.   _____

8. What is the thickness of the chassis bottom?   _____

Note: Use this diagram of a plate for problems 9 and 10.

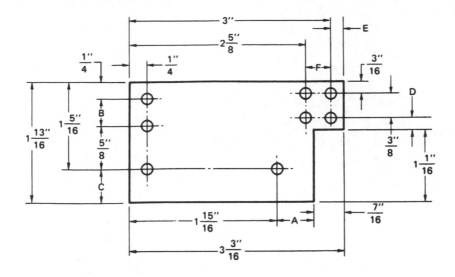

9.   Complete this table.

	DIMENSION	LENGTH
a.	A	
b.	B	
c.	C	
d.	D	
e.	E	
f.	F	

10.   The 3¾₆-inch length is changed to 3⅞ inches.

a. Find dimension A.   a. _____

b. Find dimension E.   b. _____

11.   A customer purchases 1¾ pounds of solder, ½ pound of wire, and 1⅝ pounds soldering iron. The three items are placed in a bag that can support 5 pounds. How many additional pounds can be added?

_____

12. Power resistors are mounted away from the circuit board to allow air to circulate under the resistor. How much space is allowed under the resistor in this diagram?   _____

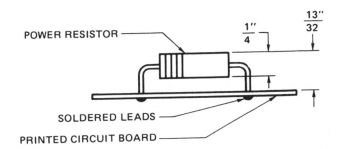

13. A delivery van has 19½ miles to travel. The first stop is made at 7¹⁄₁₀ miles, and the second stop is 4⅖ miles from the first stop. How many miles remain to complete the trip?   _____

Note: Use this diagram for problems 14 and 15.

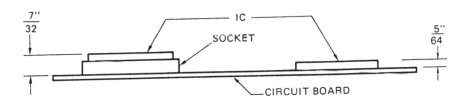

14. Integrated circuits (IC) are sometimes placed in sockets on a circuit board. Sockets are more expensive, but the IC can be removed more easily. The diagram shows an IC in a socket and one mounted directly on the board. Find the height of the socket.   _____

15. A new socket is placed on the board in the diagram. The new socket and IC height is ¼ inch instead of ⁷⁄₃₂ inch. Find the socket height.   _____

Note: Use this diagram for problems 16 and 17.

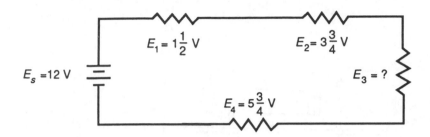

16.   The sum of voltage drops around a series circuit equals the source voltage. Find the voltage (E_3) in this circuit.

Note: $E_3 = E_s - E_1 - E_2 - E_4$

17.   A new source voltage of 18 volts is placed in the circuit in the diagram. The new voltages are $E_1 = 2\ 1/4$ volts, $E_2 = 5\frac{5}{8}$, volts, and $E_4 = 8\frac{5}{8}$, volts. Find the value of E_3.

Note: Use this circuit diagram for problems 18 and 19.

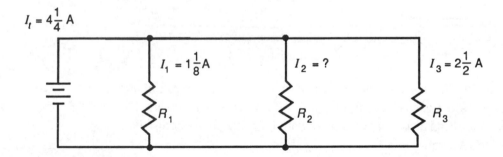

18.   The total current (I_t) in a parallel circuit is equal to the sum of the branch currents. Find the current in the second branch (I_2).

Note: $I_2 = I_t - I_1 - I_3$

19.   The total current (I_t) in the circuit is changed to $6\frac{7}{8}$ amperes. I_1 and I_3 remain the same. What is the value of I_2?

20.   A piece of solder was $42\frac{9}{10}$ centimeters long. Pieces of $6\frac{1}{2}$ centimeters, $12\frac{3}{5}$ centimeters, $4\frac{1}{5}$ centimeters, $11\frac{7}{10}$ centimeters, and $3\frac{1}{2}$ centimeters are cut off. Find the amount of solder remaining.

Unit 8   MULTIPLICATION OF COMMON FRACTIONS

BASIC PRINCIPLES OF MULTIPLYING COMMON FRACTIONS

Fractions are multiplied by simply multiplying the numerators together and the denominators together.

Example: Multiply ⅝ by ¼.

$$\frac{5}{8} \times \frac{1}{4} = \frac{5}{32}$$

When fractions are multiplied, it is often possible to simplify the problem by *cross reduction.* If a number can be found that will divide into both the numerator of one fraction and the denominator of the other, the problem can be made simpler. The answers should always be reduced to lowest terms.

Example: ⅘ × ¹⁵⁄₃₂.

$$\frac{\overset{1}{\cancel{4}}}{\underset{1}{\cancel{5}}} \times \frac{\overset{3}{\cancel{15}}}{\underset{8}{\cancel{32}}} = \frac{3}{8}$$

When a fraction is to be multiplied by a whole number, the whole number is first changed into a fraction. This is done by placing the whole number over 1.

Example: 3 × ⁷⁄₁₂

$$\frac{\overset{1}{\cancel{3}}}{1} \times \frac{7}{\underset{4}{\cancel{12}}} = \frac{1}{1} \times \frac{7}{4} = 1\frac{3}{4}$$

There are also times when it is necessary to change a *mixed* number into an *improper fraction.* A mixed number is a number that contains both a whole number and a fraction, such as 3½. To change a mixed number into an improper fraction, multiply the whole number by the denominator and then add the numerator to the result. Place the answer in the numerator of the improper fraction. Leave the denominator as it was.

Example: Change 3½ into an improper fraction

$$3\frac{1}{2} = \frac{3 \times 2 + 1}{2} = \frac{6 + 1}{2} = \frac{7}{2}$$

Example: 3⅝ × 4¼

The first step is to change both mixed numbers into improper fractions. The improper fractions can then be multiplied.

$$3\frac{5}{8} = \frac{29}{8}$$

$$4\frac{1}{4} = \frac{17}{4}$$

$$\frac{29}{8} \times \frac{17}{4} = \frac{493}{32}$$

$$\frac{493}{32} = 15\frac{13}{32}$$

PRACTICAL PROBLEMS

1.   A technician works 8¼ hours per day. How many hours are worked in 5 days?   _____

2.   One resistor costs 12½ cents. Find the cost of 24 resistors at this price.   _____

   Note: 100 cents = $1

3.   The circuit board in this diagram has 9 diodes mounted on one edge. The diodes are ⁷⁄₁₆ inch apart. Find dimension A.   _____

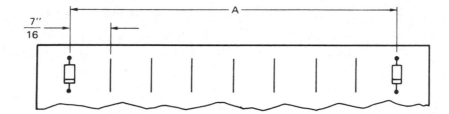

4.   A bundle contains 41 wires. Each wire is 35½ centimeters long. If the wires are soldered together, what is the total length in centimeters?   _____

5.   A stereo box is 7⅜ inches high. How high is a stack of 15 boxes?   _____

Note: The following symbols and equations are used in problems 6–9.

E = voltage in volts R = resistance in ohms

I = current in amperes P = power in watts

$$E = I \times R \qquad\qquad P = I \times E$$

6. A 5-ohm resistor (R) has a 2⅛-ampere current (I) through it. Find the voltage drop (E) across the resistor.   _____

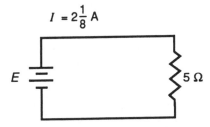

$I = 2\frac{1}{8}$ A

E

5 Ω

7. A resistance has 14¾ volts applied to it. If it draws ⅝ ampere, how many watts of power are being produced by the resistor?   _____

8. The lamp in the circuit draws ³⁄₁₀ ampere. Find its power rating in watts.   _____

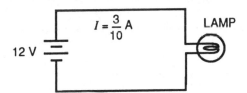

$I = \frac{3}{10}$ A   LAMP

12 V

9. A resistance draws ¹⁄₁₀ ampere. If the resistance is 680 ohms, what is the applied voltage?   _____

10. A radio kit requires a number of wires to be cut. Complete the table by filling in the length of each wire and the total length of all wires.   _____

	WIRE #	NUMBER REQUIRED	LENGTH	AMOUNT REQUIRED
a.	1	4	3½"	
b.	2	3	7¾"	
c.	3	5	4⅜"	
d.	4	1	4½"	
e.	5	3	8⁷⁄₁₆"	
			TOTAL	

11.   A delivery van travels 682 miles. The driver stops after completing ¼ of the
      trip. How many miles are traveled before stopping?   _____

12.   In alternating current, the peak voltage is ½ of the peak-to-peak voltage. The
      peak-to-peak voltage is 47¾ volts. Find the peak voltage.   _____

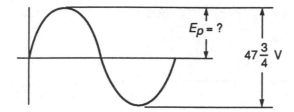

13.   A manufacturer of circuit board material cuts boards from longer pieces. Use
      this diagram and determine the length of piece needed to cut 18 boards.   _____

      Note: Assume there is no loss of board due to cutting.

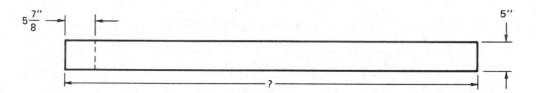

14.   A company produces 1,580 capacitors per hour. How many capacitors are
      produced in ¾ hour?   _____

15.   Each vertical centimeter on an oscilloscope shows an amount of voltage. If
      each centimeter equals 2 volts, find the peak-to-peak voltage in this diagram.   _____

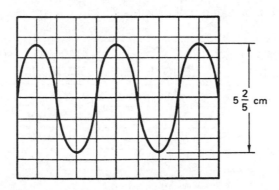

16. A radio station broadcasts its signal a distance of 480 miles from the station. About $\frac{7}{10}$ of the broadcast distance lies out of the state. How many miles of this broadcast distance are out of the state?   _____

17. A cable television tower is installed. During the first 3 days, a crew of 5 workers is used. After 3 days, 2 additional workers join this crew. The total crew works 4 more days. A workday is $8\frac{3}{4}$ hours.

    a. Find the total hours worked on the first 3 days.   a. _____

    b. Find the total hours worked over the 7 days.   b. _____

18. A supply house has a stock of 3,700 transformers. If $\frac{1}{4}$ of them are shipped out, how many transformers remain in stock?   _____

19. The power cord on a television set is $72\frac{5}{8}$ inches long. How many inches of cord are needed to equip 48 sets?   _____

20. Ferric chloride is a chemical used to etch printed circuit boards. A technician has a supply of $4\frac{3}{4}$ gallons and uses $\frac{1}{3}$ of the supply. Find the number of gallons used.   _____

Unit 9 DIVISION OF COMMON FRACTIONS

BASIC PRINCIPLES OF DIVIDING COMMON FRACTIONS

Common fractions can be divided in a manner similar to that used to multiply common fractions. When dividing common fractions, it is necessary to invert the divisor (the second fraction) and multiply. The same rules that are used for the multiplication of fractions can then be followed. All answers should be reduced to lowest terms.

Example: Divide ¾ by ⅝.

$$\frac{3}{4} \div \frac{5}{8} = \frac{3}{\cancel{4}_1} \times \frac{\cancel{8}^2}{5} = \frac{6}{5} = 1\frac{1}{5}$$

PRACTICAL PROBLEMS

1. How many 3¾-inch wires can be cut from a roll of wire 100 feet long? _____

   Note: 1 foot = 12 inches

2. A quality control tester checks out a power supply in ¾ hour. How many power supplies are tested in a 40-hour workweek? _____

3. A delivery truck travels 38¼ miles in ¾ hour. Find the speed in miles per hour. _____

4. How many complete 16⅔-centimeter test leads can be cut from a 10-meter piece of wire? _____

   Note: 1 meter = 100 centimeters

5. The transistors in the diagram are equally spaced. Find the distance between transistors. _____

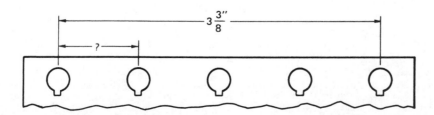

6. A set of zener diodes costs $6. Each diode costs 12½ cents. Find the number in each set.

        _____

ZENER DIODE SYMBOL

7. A roll of solder sells for $9. Each roll weighs 2¼ pounds. Find the cost per pound.

        _____

Note: The following symbols and equations are used in problems 8–12.

E = voltage in volts R = resistance in ohms

I = current in amperes P = power in watts

$$I = E \div R \quad\quad R = E \div I \quad\quad\quad I = P \div E \quad\quad E = P \div I$$

8. A load draws ³⁄₁₀ ampere. Find its resistance in ohms if it is connected to a 12-volt source.

        _____

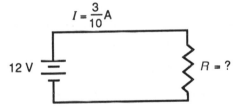

9. A load is using 38¾ watts. It operates on 10 volts. How much current (I) does it draw?

        _____

10. A rheostat is a variable resistor. The rheostat in the circuit is set to 12½ ohms. Find the value of current in this circuit.

        _____

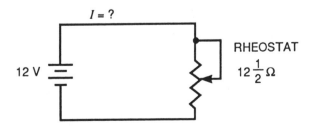

11. A resistor connected across $32\frac{1}{4}$ volts draws $1\frac{1}{2}$ amperes. What is the resistance in ohms? __________

12. The load in the schematic is dissipating $7\frac{1}{5}$ watts. Find the value of the source voltage. __________

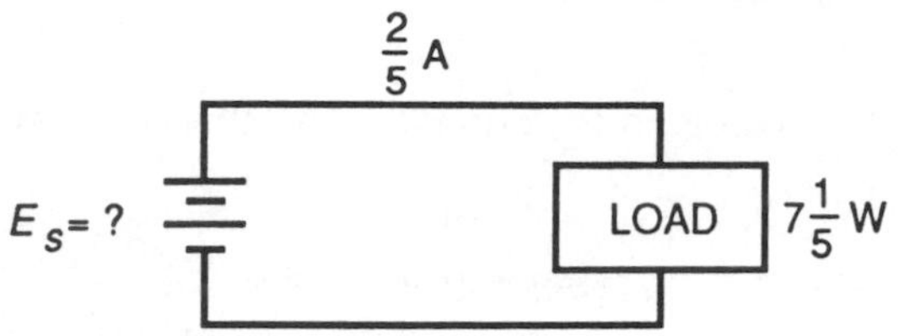

13. A record makes $16\frac{7}{8}$ revolutions in $\frac{3}{8}$ minute. Find the revolutions per minute. __________

14. A technician averages $1\frac{1}{3}$ hours per job on repair work. How many jobs can be completed during an 8-hour period? __________

15. A delivery van travels 84 miles and uses $4\frac{2}{3}$ gallons of fuel. Find the number of miles per gallon. __________

16. The current gain of an amplifier is the output current divided by the input current. The input is $\frac{3}{10}$ milliamp and the output is $7\frac{1}{2}$ milliamps. Find the current gain. __________

17. The voltage gain of an amplifier is the output voltage divided by the input voltage. The input is $\frac{3}{8}$ volt and the output is 12 volts. Find the voltage gain. __________

18. The power gain of an amplifier is the output power divided by the input power. The output is 18 watts and the input is $\frac{3}{16}$ watt. Find the power gain. __________

19. Randy is an electronics technician and is paid $847 for $38\frac{1}{2}$ hours of work. Find his salary rate per hour. __________

20. Frequency, in hertz, is the number of cycles of a signal that occur in one second. Find the frequency if 12 cycles occur in $\frac{1}{5}$ second. __________

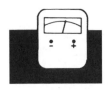

Unit 10 COMBINED OPERATIONS WITH COMMON FRACTIONS

This unit provides practical problems involving combined operations of addition, subtraction, multiplication, and division of common fractions.

PRACTICAL PROBLEMS

1. Marie drives a parts delivery truck. During a three-day period, she travels 18⅒ miles, 24⁷⁄₁₀ miles, and 21⅗ miles. She uses 4⅗ gallons of fuel. Find the number of miles per gallon.   _____

2. Drawings for a printed circuit board are sometimes drawn four times the actual size. A line on a particular drawing is 17¼ inches. What is the actual length?   _____

3. A kit requires a number of 22-gauge wires. There are eight 2¾-inch wires, three 6½-inch wires, five 1⅝-inch wires, and nine 3⅛-inch wires. The wires are cut from a piece of wire measuring 6½ feet.   _____

   a. Find in inches the total length of wire required.   a. _____

   b. Find the number of inches of wire remaining.   b. _____

   Note: The following symbols and equations are used in problems 4–10.

   E = voltage in volts R = resistance in ohms
   I = current in amperes P = power in watts
   $E = I \times R$ $P = I \times E$
   $I = E \div R$ $I = P \div E$
   $R = E \div I$ $E = P \div I$

4. A 24-ohm resistor has a voltage applied to it. Use this schematic to determine the amount of current in amperes.   _____

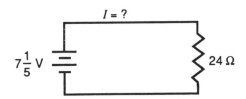

$$I = ?$$

$7\frac{1}{5}$ V   24 Ω

5. The diode in the diagram has $\frac{7}{10}$-volt drop across it. Using the current through the diode, determine its resistance.

_____

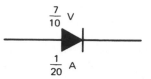

6. A load is connected to a 12-volt source. The load draws a current of $\frac{2}{3}$ ampere. Find its power rating in watts.

_____

7. A bulb is consuming $2\frac{1}{2}$ watts of electricity. Use this schematic to find how much current it is drawing.

_____

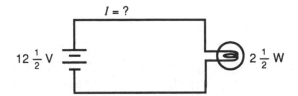

8. The values of resistors in series are added to form total resistance (R_t). If $\frac{3}{4}$ ampere is carried through the circuit, find the source voltage.

    Note: $R_t = R_1 + R_2 + R_3 + R_4$

    $E_s = I \times R_t$

_____

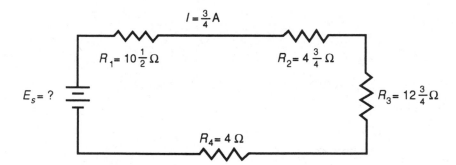

9. A resistor in a CD player is rated at 68 ohms. While troubleshooting, a technician measures $10\frac{1}{5}$ volts across the resistor. Find the current flow through the resistor.

_____

10. Find the resistance of the load in the circuit when 6³⁄₁₀ volts are applied. _____

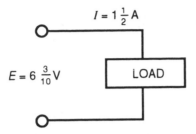

$$I = 1\frac{1}{2}\,A$$

$$E = 6\frac{3}{10}\,V$$

LOAD

11. A cable TV installer earns $16 per hour and works 38½ hours per week. Deductions for taxes and a savings account deposit are ⅓ of the weekly pay. How much remains of the weekly earnings after deductions are made? _____

12. A roll of wire weighs 4 pounds and costs $12. What is the cost of ⅓ pound? _____

13. Rick is an assembler in an electronics factory. It takes Rick ¾ hour to assemble a radar detector. He is paid $9. Find his total pay for 19 ¼ hours. _____

14. A shop receives 18, 27, and 48 cassette tape recorders in three shipments. During a sale, ⅓ of them are sold for $28 each. The rest are sold for $32 each. Find the total income. _____

15. A motor makes 38¾ revolutions in ¼ of a second. How many revolutions are completed in one second? _____

16. Light travels 186,000 miles per second. Find the number of miles traveled in ³⁄₂₀ second. _____

17. A container of printed circuit board cleaner contains 1½ gallons. During a one-month period, 1⅛ pints, 2¼ pints, ¾ pint, 3⅜ pints, and ⅞ pint are used. How many pints remain in the container? _____

   Note: 1 gallon = 8 pints

18. During a contest, a tape is played every 3½ minutes. How many tapes are played during 1 hour 10 minutes? _____

19. Sound travels at a speed of 1,100 feet per second. How many feet does it travel in ¹¹⁄₂₀ second? _____

20. An electronics book has 496 pages. If a student reads 186 pages, what fraction of the total pages is read?   _____

21. An ampere is defined as an electron flow rate of one coulomb per second. If 3½ coulombs flow through a 220-ohm resistor in 8¾ seconds, how much voltage is dropped across the resistor?   _____

    Note: $E = I \times R$

22. A box weighs 1¾ pounds. A cassette tape weighs ¼ pound. The total weight of the box and tapes is 64¼ pounds. How many tapes are in the box?   _____

Decimal Fractions

Unit 11   ROUNDING NUMBERS

BASIC PRINCIPLES OF ROUNDING NUMBERS

In electronics, a calculator is often used to solve equations, formulas, and other practical problems. A calculator will usually provide an answer that is carried out to more decimal places than the technician requires. When this occurs, the technician must *round* the number to the required number of decimal places. Before rounding a number, the technician must first be familiar with the name and value of each decimal place. These are:

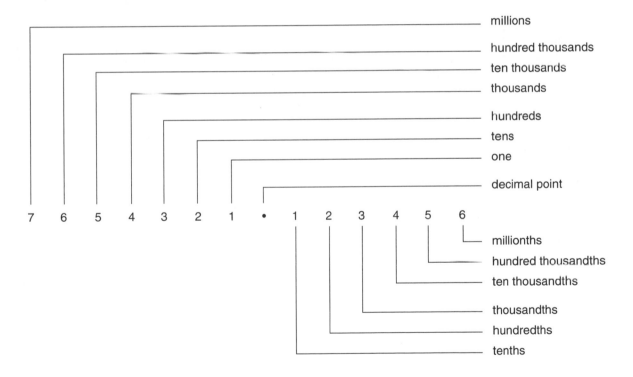

In order to round a number to a given decimal place, follow these rules:

1. Locate the place (value) to be rounded and look at the number immediately to the right of this number.

2. If the number occupying this place is 4 or less, the number to be rounded remains the same and all numbers to the right of this number are removed (for numbers to the right of the decimal point) or changed to zeroes (for numbers to the left of the decimal point).

3. If the number occupying this place is 5 or more, the number to be rounded is increased by 1 and all numbers to the right of this number are removed or changed to zeroes, as explained above.

Example: Round 48,637.02 to the nearest thousand.

The number occupying the thousands place is 8. The number to the right of 8 is 6. Since 6 is greater than 4, the 8 is increased to 9 and all other numbers are changed to zeroes. The result is 49,000.

Example: Round 0.085 407 to the nearest thousandth.

The number occupying the thousandths place is 5. Since the number to the right of the 5 is a 4, the final answer is 0.085.

PRACTICAL PROBLEMS

In problems 1–10, round the number to the number of decimal places indicated in parentheses.

1. 12.878 V (2) =   _____

2. 0.003 499 A (3) =   _____

3. 485.627 Ω (1) =   _____

4. 9.682 351 V (4) =   _____

5. 1,414.836 51 Hz (3) =   _____

6. 0.000 997 089 F (6) =   _____

7. 0.240 654 A (5) =   _____

8. 956.005 Ω (2) =   _____

9. 0.909 A (1) =   _____

10. 90.416 V (3) =   _____

11. A 2.7 kilohm resistor is determined to have a resistance of 2,755 ohms. Round this measured value to the nearest hundred ohms.   _____

12. A digital voltmeter displays a voltage of 118.145 volts. Round this value to the nearest volt.   _____

13. Round the current value flowing through R_1 in this circuit to the nearest tenth of a milliampere (mA).   _____

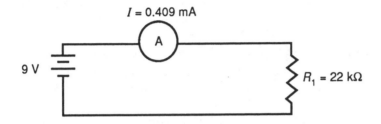

$I = 0.409$ mA

A

9 V

$R_1 = 22$ kΩ

14. The value of R_1 in problem 13 is 22,000 ohms. Round this value to the nearest ten thousand ohms.   _____

15. A frequency counter displays a value of 153,725 hertz. Round this frequency to the nearest ten hertz.   _____

16. The actual value of capacitance in a circuit is 0.000048 farad. What is this value rounded to five decimal places?   _____

17. A calculator is used to solve for current in a circuit. The calculated value is 0.087 254 ampere. Round this to the nearest thousandth of an ampere.   _____

18. The total impedance of a circuit is calculated at 48,725.36 ohms. Round this value to the nearest hundred ohms.   _____

19. An extremely low current is detected in a circuit. Round this current of 0.000 003 516 ampere to the nearest millionth of an ampere.   _____

20. Round the current in problem 19 to the seventh decimal place.   _____

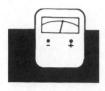

Unit 12   CONVERSION OF COMMON AND DECIMAL FRACTIONS

BASIC PRINCIPLES OF CONVERTING FRACTIONS

There are times when it is necessary or desirable to change common fractions into their decimal equivalent, or to change a decimal fraction into a common fraction. The following procedures can be used to achieve these conversions.

A common fraction is actually an algebraic expression of division. In algebra, any time one number is shown above another, it means to divide the top number by the bottom number. Therefore, to convert any common fraction into its decimal equivalent, divide the numerator by the denominator.

Example: Convert the common fraction ¾ to its decimal equivalent.

$$\frac{3 \text{ (numerator)}}{4 \text{ (denominator)}}$$

$$\frac{3}{4} = 4\overline{)3} = 0.75$$

A decimal fraction can be changed to a common fraction, as shown in the following example.

Example: Change 0.35 to a common fraction.

The first step is to determine the denominator. To do this, name the place values of the decimal fraction. The first number to the right of the decimal point is in the *tenths* position, the second number is in the *hundredths* position, the third number is in the *thousandths* position, and so on. In the fraction 0.35, the 5 is in the hundredths position. Therefore, 0.35 can be rewritten as 35 hundredths.

$$\frac{35}{100}$$

The next step is to reduce the fraction to lowest terms. In this example, both the numerator and denominator are divisible by 5.

$$\frac{35}{100} = \frac{7}{20}$$

Example: Convert 0.250 into a common fraction.

In this example, the 0 to the right of the 5 is in the thousandths position. Rewrite the decimal fraction as the common fraction $^{250}/_{1000}$.

$$\frac{250}{1000}$$

Reduce the fraction to lowest terms.

$$\frac{250}{1000} = \frac{1}{4}$$

PRACTICAL PROBLEMS

Note: Reduce all fractions to lowest terms.

1.   Express the decimal fraction 12.25 as a common fraction.   _____

2.   Express the decimal fraction 1.1 as a common fraction.   _____

3.   Express the decimal fraction 19.02 as a common fraction.   _____

4.   Express the decimal fraction 3.55 as a common fraction.   _____

5.   Express the decimal fraction 5.875 as a common fraction.   _____

6.   Express the decimal fraction 1.015625 as a common fraction.   _____

7.   Find the length of the integrated circuit in this diagram expressed as a decimal fraction.   _____

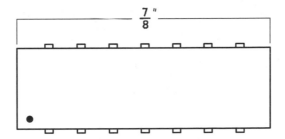

8.   The rating of a resistor is 1.75 watts. Express this rating as a mixed number.   _____

9.   A piece of cable is 7$^{3}/_{32}$ inches long. Express this length as a decimal.   _____

10. A section of copper wire has a resistance of 10.15 ohms. Express this resistance as a mixed number.

_____

11. Number ten copper wire is 0.102 inch in diameter. Find this diameter if expressed as a fraction.

_____

12. The current gain of a common base amplifier is 0.84. Write this as a fraction.

_____

13. The meter in this circuit indicates the amount of current. Express this as a decimal.

_____

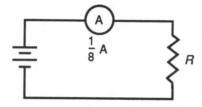

14. The conductance of a resistor is found by dividing 1 by the resistor value. Find the conductance (G) of a 25-ohm resistor (R).

_____

Note: $G = \frac{1}{R}$

15. The sine of a particular angle is 0.5. Write this as a fraction.

_____

16. The square root of 5.569 6 is 2.36. Express this square root as a mixed number.

_____

17. A technician repairs 91 out of 104 computers during a six-month period. Written as a fraction, this would be $\frac{91}{104}$. Write this as a decimal.

_____

18. A photovoltaic cell produces voltage when it is struck by light. Write the voltage output of this cell as a decimal.

_____

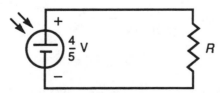

19.   A digital voltmeter indicates a voltage of 0.35 volt. Express this as a fraction.   _____

20.   A fuse is rated at ¹⁄₁₆ ampere. Express this as a decimal.   _____

Unit 13 ADDITION OF DECIMAL FRACTIONS

BASIC PRINCIPLES OF ADDING DECIMAL FRACTIONS

One of the advantages of decimal fractions as compared to common fractions is that when decimal fractions are added or subtracted, it is not necessary to find a common denominator. Adding decimal fractions is accomplished by placing the fractions in a column with all the decimal points aligned. The columns are then added in the same manner as are whole numbers.

Example: 4.0563 + 0.98 + 14.0008 + 0.4005 + 10.33

$$
\begin{array}{r}
4.0563 \\
0.9800 \\
14.0008 \\
0.4005 \\
+\ 10.3300 \\
\hline
29.7676
\end{array}
$$

PRACTICAL PROBLEMS

1. Find the overall length of the potentiometer in the diagram. _____

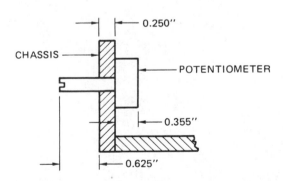

2. A technician has 6 pieces of solder. The lengths are 9.2 centimeters, 15.84 centimeters, 12.7 centimeters, 14.68 centimeters, 10.1 centimeters, and 18.4 centimeters. Find the total length of the six pieces. _____

3. The values of voltages in series are added to find total voltage. Voltages of 17.5 volts, 24.12 volts, 12.8 volts, and 18.55 volts are in series. Find the total voltage. _____

4. An electronic mixer can add the two frequencies put into it. Frequencies of 37.128 kilohertz and 65.289 kilohertz are mixed. What is the output sum? _____

Note: Use this diagram for problems 5–7.

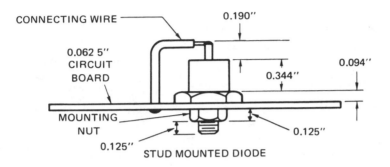

5. Find the total length the diode extends below the board. _____

6. Find the total length the diode extends above the surface of the board. _____

7. Find the overall length of the diode. _____

8. The time clock at a plant measures time to the nearer hundredth hour. During a five-day workweek, Jeff works 8.15 hours, 8.24 hours, 8.76 hours, 9.03 hours, and 8.75 hours. What is the total number of hours he works? _____

9. The resistor in this diagram is bent to fit into a circuit board. What is the total length of the resistor and the leads? _____

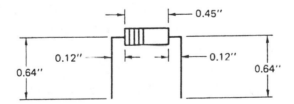

10.   The values of branch currents in a parallel circuit are added to form total current (I_t). Find total current in this circuit.   _____

Note: $I_t = I_1 + I_2 + I_3$

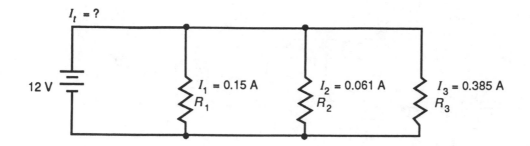

11.   A liquid developer is used in making circuit boards. A solution is prepared using 1.135 liters of developer and 5.675 liters of water. Find the number of liters contained in the solution.   _____

12.   Jerry purchases several items at an electronics supply store. Find the total cost of these items.   _____

ITEM	AMOUNT
Resistors	$ 1.49
Fuses	$ 0.84
Soldering Iron	$ 15.95
Solder	$ 2.65
Wire	$ 4.50

13.   A van travels 19.8 miles, 14.7 miles, and 36.28 miles to make deliveries. From the last delivery point, the distance back to the warehouse is 24.8 miles. Find the total distance driven.   _____

14. The values of resistors connected in series are added to form total resistance (R_t). Find total resistance in this circuit. _____

    Note: $R_t = R_1 + R_2 + R_3 + R_4$

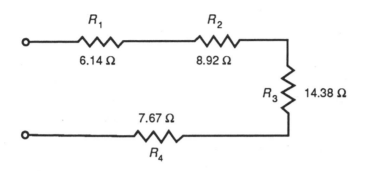

15. A repair technician charges $86.56 for parts and $147.30 for labor. The customer also orders another $65.00 in replacement parts. What is the total charge to the customer? _____

16. The values of inductors connected in series are added to form total inductance (L_t). Find L_t in the circuit shown. _____

    Note: $L_t = L_1 + L_2 + L_3 + L_4$

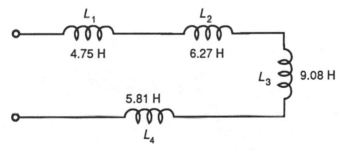

17. A set of relay contacts are 0.028 7 inch apart. They are moved an additional 0.003 6 inch apart. What is the total distance between the contacts? _____

18. Peak-to-peak voltage is the sum of both peaks of a signal. Find the peak-to-peak voltage in the diagram.   _____

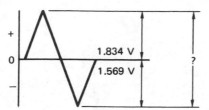

19. A common emitter amplifier has an input resistance of 925.76 ohms. This resistance is increased by 235.38 ohms. What is the total resistance?   _____

20. The emitter current in a transistor is equal to the base current plus the collector current. Find emitter current (I_e) in this circuit.   _____

    Note: $I_e = I_b + I_c$

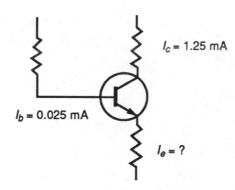

Unit 14   SUBTRACTION OF DECIMAL FRACTIONS

BASIC PRINCIPLES OF SUBTRACTING DECIMAL FRACTIONS

When decimal fractions are subtracted, place the smaller number below the larger number taking care to keep the decimal points aligned. The procedure is then the same as subtraction of whole numbers.

Example: 12.349 − 8.907

$$
\begin{array}{r}
12.349 \\
- \ 8.907 \\
\hline
3.442
\end{array}
$$

PRACTICAL PROBLEMS

1. A piece of circuit board 7.75 inches long is cut from the larger piece. Find the length of the remaining piece.

_____

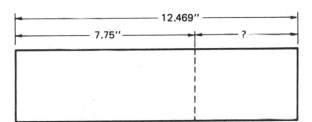

2. Randy has a balance of $676.87 in his checking account. He buys a stereo for $289.00, two speakers for $87.50 each, and a set of headphones for $39.95. How much money remains in his account?

_____

3. An electronics assembly kit includes a piece of wire 3.8 feet long. Pieces 0.56 feet, 0.87 feet, 0.12 feet, 0.08 feet, 0.06 feet, and 0.58 feet are cut off. Find the length of the remaining piece.

_____

4.   Circuit breakers are used to protect equipment from receiving too much current. The breaker in this schematic opens at 5 amperes of current. How many more amperes of current will cause the breaker to open?   _____

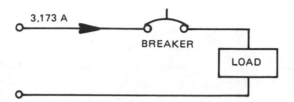

5.   A piece of metal 37.75 inches long is used to make some chassis plates. Pieces 12.625 inches, 6.437 5 inches, and 14.875 inches are cut. Find the remaining length.   _____

6.   In a transistor, the emitter current (I_e) is equal to base current (I_b) plus collector current (I_c). Calculate the collector current in the circuit.   _____

Note: $I_e = I_b + I_c$

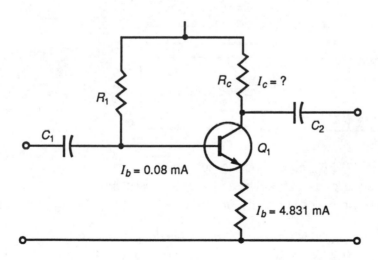

Note: Use this diagram for problems 7–9.

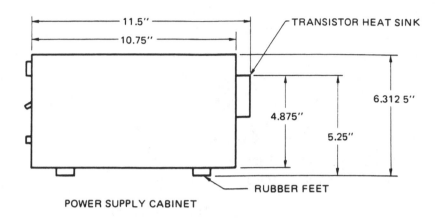

POWER SUPPLY CABINET

7.   How far does the heat sink extend beyond the back of the cabinet?   _____

8.   How high are the rubber feet?   _____

9.   What is the distance from the top of the cabinet to the top of the heat sink?   _____

Note: Use this formula and diagram for problems 10 and 11.

$$E_{R_3} = E_s - E_{R_1} - E_{R_2} - E_{R_4}$$

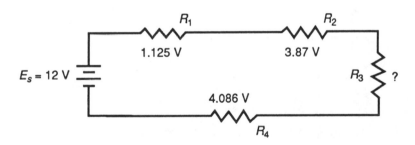

10.   Voltage drops in a series circuit add up to total supply voltage. If the supply voltage in the circuit is 12 volts, find the voltage across R_3.   _____

11.   A new supply voltage of 30 volts is placed in the circuit. The voltage drops are now $E_{R_1} = 2.812\ 5$ volts, $E_{R_2} = 9.675$ volts, $E_{R_4} = 10.215$ volts. Find the voltage across R_3.   _____

Note: Use this formula and diagram for problems 12 and 13.

$$I_{R_1} = I_t - I_{R_2} - I_{R_3} - I_{R_4}$$

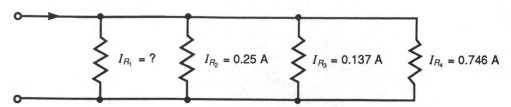

$I_{R_1} = ?$ $I_{R_2} = 0.25$ A $I_{R_3} = 0.137$ A $I_{R_4} = 0.746$ A

12.   Branch currents in a parallel circuit add up to total supply current (I_t). What is the current through R_1 if the total circuit current is 1.836 amperes?   _____

13.   A total current flow of 1.377 amperes enters the circuit. The branch currents become $I_{R_2} = 0.187\,5$ amperes, $I_{R_3} = 0.102\,75$ amperes, and $I_{R_4} = 0.559\,5$ amperes. Find I_{R_1}.   _____

14.   One roll of solder weighs 4.25 pounds, and the other roll weighs 5.18 pounds. Each roll costs $42.65. How many more pounds does the heavier roll weigh?   _____

15.   The peak-to-peak voltage of a square wave is 12.962 volts. Use this diagram to determine the value of the upper portion of the wave.   _____

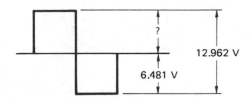

?

12.962 V

6.481 V

16.   An electronic mixer can produce the difference in two input frequencies. Find the difference if 74.325 megahertz and 48.723 megahertz are mixed.   _____

17.   Find the length of the tip of the banana plug in this diagram.   _____

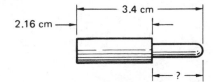

2.16 cm   3.4 cm   ?

18. One type of solder melts at 363.14°F. A soldering iron is heated to 357.96°F. How many more degrees must the iron be heated to melt the solder?   _____

19. A repair technician charges $87.50 to repair a piece of equipment. Costs include $12.55 for a transformer, $3.90 for a power transistor, and $6.00 for other parts. How much profit is made?   _____

20. The values of resistors connected in series are added to form total resistance (R_t). The total resistance in this circuit is 251.53 ohms. Find the value of R_2.   _____

   Note: $R_2 = R_t - R_1 - R_3 - R_4 - R_5$

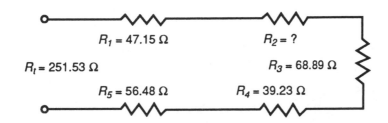

$R_1 = 47.15\ \Omega$ $R_2 = ?$

$R_t = 251.53\ \Omega$ $R_3 = 68.89\ \Omega$

$R_5 = 56.48\ \Omega$ $R_4 = 39.23\ \Omega$

Unit 15   MULTIPLICATION OF DECIMAL FRACTIONS

BASIC PRINCIPLES OF MULTIPLYING DECIMAL FRACTIONS

The multiplication of decimal fractions follows the same basic procedure as the multiplication of whole numbers. When decimal fractions are multiplied, the number of places to the right of the decimal point in both numbers multiplied must be counted. The same number of places must appear to the right of the decimal point in the answer.

Example: 8.650×3.5

$$
\begin{array}{r}
8.650 \text{ (Three decimal places)} \\
\times \quad 3.5 \text{ (One decimal place)} \\
\hline
4\ 3250 \\
25\ 950 \\
\hline
30.2750 \text{ (Four decimal places)}
\end{array}
$$

PRACTICAL PROBLEMS

1. A technician for a computer company works 76.5 hours during a two-week period. What is the total pay if the hourly rate is $17.50?   _____

2. The rms voltage indicates the DC equivalent of the AC voltage. The rms voltage is found by multiplying 0.707 times the peak AC voltage. Find the rms voltage in this diagram.   _____

    Note: $E_{rms} = 0.707 \times E_p$

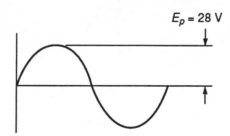

$E_p = 28$ V

3. An AC signal has a peak voltage (E_p) of 167 volts. Find the rms voltage.   _____

4. Multiply rms voltage by 1.414 to find peak voltage. If an rms voltage has a value of 40 volts, find the peak value. _____

    Note: $E_p = 1.414 \times E_{rms}$

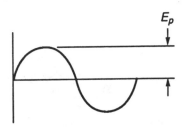

5. Find the peak value of an AC voltage with an rms value of 120 volts. _____

6. To express peak-to-peak voltage as rms voltage, multiply the peak-to-peak value by 0.353 5. Find the rms voltage of the signal in this diagram. _____

    Note: $E_{rms} = 0.353\ 5 \times E_{pp}$

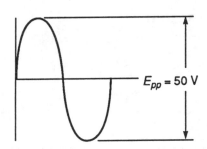

$E_{pp} = 50$ V

7. Find the rms value of an AC voltage with a peak-to-peak voltage of 622 volts. _____

8. The output voltage of an amplifier is found by multiplying input voltage by the amplifier's voltage gain. Find the output voltage of the amplifier in the diagram which shows a gain of 10.1. _____

9. A delivery van travels 48 miles per hour for 2.75 hours. How many miles does
    it travel?

    _____

10. A television station broadcasts to an area in a circle around the transmitter.
    The radius (*r*) of this broadcast circle is 48 miles. Find the circumference (*C*)
    of the circle.

    Note: $C = 2\pi r$ where $\pi = 3.1416$

    _____

11. An oscillator has an output of 4,500 cycles per second (Hz). How many
    cycles does it generate in 0.35 second?

    _____

Note: These symbols and equations are used in problems 12–15.

E = voltage in volts	*R* = resistance in ohms
I = current in amperes	*P* = power in watts
$E = I \times R$	$P = I \times E$

12. A resistor has a 0.85-ampere current through it. Use this schematic to
    determine the applied voltage.

    _____

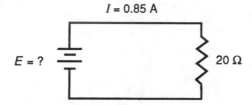

$$I = 0.85 \text{ A}$$

$$E = ?$$ $$20\ \Omega$$

13. A 48-ohm resistor is placed across 6 volts. The current is 0.125 ampere. Find
    the power in watts consumed by the resistor.

    _____

14. What is the value of the source voltage in this circuit?

    _____

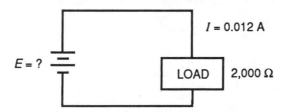

$$I = 0.012 \text{ A}$$

$$E = ?$$ LOAD $$2,000\ \Omega$$

15. A 7,200-ohm resistor has 0.005 ampere of current flowing through it.

    _____

    a. Find the voltage drop across the resistor.   a. _____

    b. Find the power consumed by the resistor.   b. _____

16. A supply store purchases 150 tuning capacitors. Each capacitor costs $1.27. How much is the total purchase?   _____

TUNING CAPACITOR   GANGED TUNING CAPACITOR

17. Radio signals travel at 186,000 miles per second. How far does a signal travel in 0.04 second?   _____

18. The output current (I_c) of the transistor amplifier is the input current (I_b) multiplied by the current gain. Find I_c if the amplifier gain is 70.   _____

   Note: $I_c = I_b \times$ gain

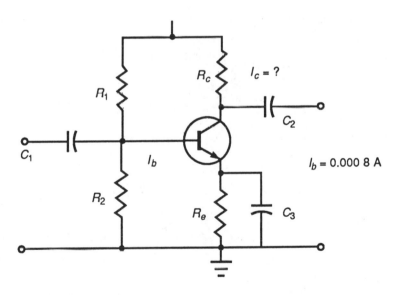

19. The input voltage of an amplifier is 0.4 volt and the gain is 25. Find the output voltage.   _____

   Note: $E_{out} = E_{in} \times$ gain

20. Marlene delivers parts for a supply store. She receives a salary plus 21¢ per mile for the use of her car. One delivery takes 2.5 hours to complete. She averages 52 miles per hour. How much does she receive for the use of her car on this delivery?   _____

Unit 16   DIVISION OF DECIMAL FRACTIONS

BASIC PRINCIPLES OF DIVIDING DECIMAL FRACTIONS

When decimal fractions are divided, the divisor is placed to the left of the dividend in the same manner as in the division of whole numbers. When dividing decimal fractions, however, the divisor must be a whole number and not a fraction. The divisor can be made a whole number by moving the decimal point all the way to the right of the number. When this is done, the decimal point of the dividend must be moved the same number of places to the right. The location of the decimal point in the quotient will be directly above the decimal point in the dividend. The numbers are then divided in the same manner as are whole numbers.

Examples: Divide the quantities in the following example problems.

a.   2.67 ÷ 0.3 ⟶

```
            8 .9
   0 3. ) 2 6 .7
           2 4
           2 7
           2 7
             0
```

b.   329.868 ÷ 7.14 ⟶

```
              4 6 . 2
   7 1 4. ) 3 2 9 8 6 . 8
             2 8 5 6
               4 4 2 6
               4 2 8 4
                 1 4 2  8
                 1 4 2  8
                      0
```

c.   100 ÷ 4.4 ⟶

```
              2 2 . 7 2 7
   4 4. ) 1 0 0 0 . 0 0
           8 8
           1 2 0
             8 8
             3 2  0
             3 0  8
               1  2 0
                  8 8
                  3 2 0
                  3 0 8
```

PRACTICAL PROBLEMS

1. The quality (Q) of a tuned circuit is the inductive reactance (X_L) divided by resistance (R). Find the quality of this circuit.

    Note: $Q = X_L \div R$

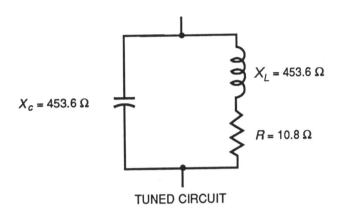

$X_L = 453.6 \ \Omega$

$X_c = 453.6 \ \Omega$

$R = 10.8 \ \Omega$

TUNED CIRCUIT

2. A roll of wire is 250 feet long. How many 3.25-foot wires can be cut from the roll?

3. A disc jockey plays 8 songs in 0.75 hour. How many songs are played in 3 hours?

4. A supplier buys a box of special integrated circuits for $85.50. Each IC costs $4.75. Find the total number in the box.

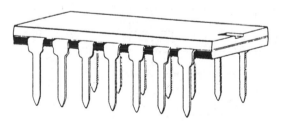

Adapted with permission from Richard A. Honeycutt, *Op Amps and Linear Integrated Circuits*.
Copyright 1988 by Delmar Publishers Inc, p. 6.

5. A television signal travels 22,320 miles in 0.12 second. Find the speed in miles per second.

6. The period (T) of an AC signal is the time in seconds that it takes one cycle to occur. The period is found by dividing 1 by the frequency (f). Find the period of the signal in this diagram.

   Note: $T = 1 \div f$ or $\frac{1}{f}$

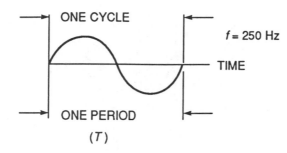

ONE CYCLE

$f = 250$ Hz

TIME

ONE PERIOD

(T)

7. Find the period (T) in seconds of a signal with a frequency (f) of 60 hertz. Round the answer to four decimal places.

8. The frequency (f) in hertz of a signal can be found by dividing 1 by the period (T). If the period is 0.002 5 second, find the frequency.

   Note: $f = 1 \div T$ or $\frac{1}{T}$

9. A signal is shown on the screen of an oscilloscope. Based upon scope settings, its period is determined. Use the period (T) in the diagram and find the signal frequency (f).

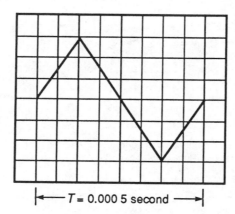

$T = 0.000\ 5$ second

10. A supply van travels 68 miles in 1.25 hours. What is the speed in miles per hour?

11. A 1,000-foot roll of wire sells for $23.10. Find the cost per pound if the roll weighs 6 pounds. _____

12. The four integrated circuits in the diagram are equally spaced on a circuit board. What is the distance between each circuit? _____

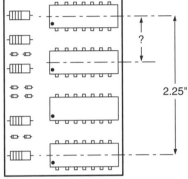

2.25"

13. It takes a cable television installer 1.2 hours to run wires to connect a house. How many installations can be made in 8.4 hours? _____

14. The current gain of an amplifier is the output current divided by the input current. The input current is 0.000 8 ampere and the output 0.072 ampere. Find the amplifier gain. _____

Note: These symbols and equations are used in problems 15–20.

E = voltage in volts $\qquad$ R = resistance in ohms

I = current in amperes $\qquad$ P = power in watts

$I = E \div R \qquad R = E \div I \qquad I = P \div E \qquad E = P \div I$

15. A resistor has a voltage drop of 3.6 volts across it. Use the diagram to calculate its resistance in ohms. _____

$I = 0.045$ A

16. A 470-ohm resistor has a voltage drop of 7.05 volts across it. Find the current in amperes. _____

17.   Using the schematic, determine the value of the source voltage (E_s).   _____

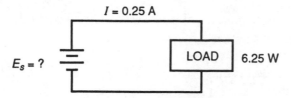

18.   A 100-ohm load is connected to a 12-volt battery. Find the current in amperes.   _____

19.   Using this schematic, determine the current passing through the load.   _____

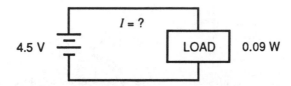

20.   A resistor draws 0.05 ampere of current when connected to 12 volts. Find the resistance in ohms.   _____

Unit 17   COMBINED OPERATIONS WITH DECIMAL FRACTIONS

This unit provides practical problems involving combined operations of addition, subtraction, multiplication, and division of decimal fractions.

PRACTICAL PROBLEMS

1. A technician works 37.5 hours and earns $18.75 per hour. Payroll deductions total $158.78. Find the total take-home pay.   _____

2. A supplier pays $116.64 for a carton of 36 field effect transistors. What is the cost of each transistor?   _____

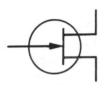

N-CHANNEL FET

P-CHANNEL FET

3. The voltage gain of an amplifier is 48.8. This means that output voltage is 48.8 times the input voltage. Find the output voltage if the input voltage is 0.07 volt.   _____

4. Five wires are connected together. Find the total length if the wires are 11.3 inches, 19.7 inches, 14.75 inches, 6.15 inches, and 18.04 inches long.   _____

5. Fred had $1,350 to invest in audio equipment. Use this bill and determine how much money he has left.   _____

ITEM	COST
Tuner	$395.99
Amplifier	287.50
CD player	237.95
Speakers	165.00
Headphones	35.95
TOTAL	

6. The current gain of an amplifier is the output current divided by the input current. Find the gain if the input current is 0.000 4 ampere and the output current is 0.068 ampere.

_____

7. The frequency (f) in hertz is 1 divided by the period (T) of that signal. Find the frequency of the signal in this diagram.

_____

   Note: $f = 1 \div T$ or $\frac{1}{T}$

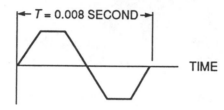

Note: These symbols and equations are used in problems 8–11.

$$E_{rms} = 0.707 \times E_p$$

$$E_{rms} = 0.353\,5 \times E_{pp}$$

$$E_p = 0.5 \times E_{pp}$$

$$E_p = 1.414 \times E_{rms}$$

$$E_{pp} = 2.828 \times E_{rms}$$

where E_{rms} = rms voltage
   E_p = peak voltage
   E_{pp} = peak-to-peak voltage

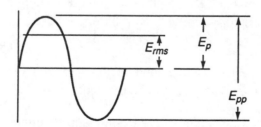

8. Find the rms voltage if a signal has a peak-to-peak voltage of 12 volts.

_____

9. An oscilloscope shows the peak voltage of a sine wave to be 24 volts. What is the rms value?

_____

10. What is the peak-to-peak voltage of a signal with an rms value of 36 volts?

_____

11. Using this diagram of a signal, find these values.

a. E_{pp}

b. E_p

c. E_{rms}

a. _____

b. _____

c. _____

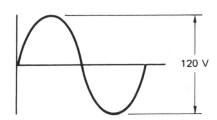

120 V

Note: These symbols and equations are used in problems 12–20.

E = voltage in volts R = resistance in ohms

I = current in amperes P = power in watts

$$E = I \times R$$
$$I = E \div R$$
$$R = E \div I$$

$$P = I \times E$$
$$I = P \div E$$
$$E = P \div I$$

12. Find the voltage drop across the resistor in this diagram.   _____

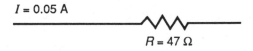

$I = 0.05$ A

$R = 47\ \Omega$

13. A load connected to 12 volts draws a current of 1.5 amperes. What is the load
     resistance in ohms?   _____

14. Calculate the source voltage (E_s) in this schematic.   _____

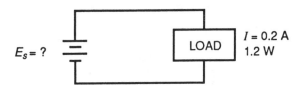

$E_s = ?$ LOAD $I = 0.2$ A
                                 1.2 W

15. A 5,600-ohm resistor has a 14-volt drop across it. Find the current in
     amperes.   _____

16. The voltage across a power transistor is 6.5 volts. Use this diagram and determine the power consumption in watts.

_____

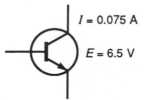

$I = 0.075$ A

$E = 6.5$ V

17. A load consumes 4.8 watts when connected to 12 volts. What is the current in amperes?

_____

18. Calculate the resistance of the light-emitting diode (LED) in this diagram.

_____

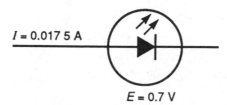

$I = 0.017\,5$ A

$E = 0.7$ V

19. A 100-ohm resistor is connected across 24 volts. The current drawn by the resistor is 0.24 ampere. Find the power in watts consumed by the resistor.

_____

20. The solar cell in the circuit is connected to a 680-ohm load. Using the circuit current of 0.002 ampere, calculate the voltage output of the cell.

_____

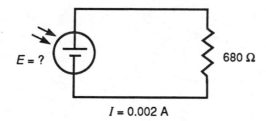

$E = ?$ 680 Ω

$I = 0.002$ A

21. A building is 88.6 feet in height. An antenna pole mounted on top of the building is 12.75 feet in height. The antenna is mounted 0.5 foot below the top of the building. What is the height of the antenna above the ground?   _____

22. A number 12 AWG (American Wire Gage) copper conductor has a resistance of 1.588 ohms per 1,000 feet at a temperature of 20° C. An electric motor is located 110 feet from its source of power. The starting current of the motor is 15.75 amperes. What is the voltage drop across the conductors when the motor starts?   _____

    Note: Conductor length must be doubled.

Exponents, Electronics Units, and Roots

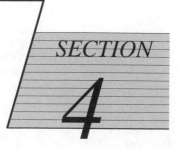

Unit 18   EXPONENTS

BASIC PRINCIPLES OF POWERS

The *power* of a number is also known as its *exponent.* If a number is *squared,* for example, it is raised to the second power, or has an exponent of 2. The exponent is the small number written above and to the right of the number as shown. This means that the number is to be multiplied by itself.

$$5^2 = (5 \times 5) = 25$$

Notice that 5 is not multiplied by 2, but is multiplied by itself.

If a number is *cubed,* it means that the number is raised to the third power, or has an exponent of 3.

$$8^3 = (8 \times 8 \times 8) = 512$$

A number can be raised to any power by using it as a factor that number of times.

Example: Raise 6 to the fifth power.

$$6^5 = (6 \times 6 \times 6 \times 6 \times 6) = 7,776$$

On a calculator, this would be entered:

$$\boxed{6} \;\; \boxed{y^x} \;\; \boxed{5} \;\; \boxed{=}$$

PRACTICAL PROBLEMS

1. $10^6 =$ _____

2. $0.01^2 =$ _____

3. $24^2 =$ _____

4. $(\frac{2}{5})^4 =$ _____

5. $9.7^1 =$ _____

6. $1.01^2 =$ _____

7. $(4\frac{1}{8})^2 =$ _____

8. $3.3^3 =$ _____

9. $19^4 =$ _____

10. $20^4 =$ _____

11. A square printed circuit board measures 4.5 inches on each side. The area
    (A) equals the length of one side (s) squared. Find the area in square inches. _____

    Note: $A = s^2$

Note: Use this information for problems 12–15.

Most wires used in electronics work have circular cross sections and have small diameters. The *mil,* a
unit used to measure these small diameters, is a unit of length equal to 0.001 inch. The first circle shown
represents the end of a piece of wire that is 0.001 inch in diameter. The cross-sectional area of this wire
is one *circular mil* (CM). To find the cross-sectional area in circular mils, square the diameter in mils.

$$A = d^2$$

The second circle is 2 mils in diameter and 4 CM in cross-sectional area because $d = 0.002$ inch,
therefore

$$A = d_2 = (2\text{mils})^2 = 4\text{ CM}$$

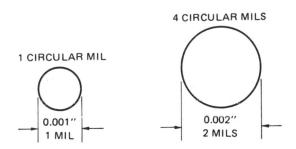

4 CIRCULAR MILS

1 CIRCULAR MIL

0.001″
1 MIL

0.002″
2 MILS

12. Find the circular mil area of a wire that is 80.8 mils in diameter. _____

13. Use this diagram of the wire to find the circular mil area. _____

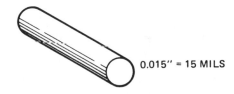

0.015″ = 15 MILS

14. A piece of 22-gauge wire has a diameter of 25.3 mils. What is the circular mil area?

_____

15. A multi-strand cable contains 20 strands of wire. Each strand is 12.6 mils in diameter. Find the total circular mil area.

_____

16. A tank is used to etch printed circuit boards. The tank is a cube with each side 1½ feet long. Find the volume (*V*) in cubic inches.

    Note: $V = s^3$

_____

17. A computer disk storage cabinet is shaped like a cube. Use this diagram to determine the volume in cubic centimeters.

_____

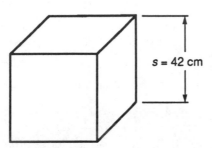

s = 42 cm

18. A resistor is stamped 56×10^3 ohms. How many ohms is this equivalent to?

_____

19. The circuit board in the diagram is square. How many square inches of board are needed to make 80 boards?

    Note: $A = s_2$

_____

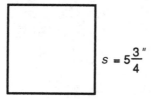

$s = 5\frac{3}{4}''$

20. The actual circuit area of an integrated circuit is square. Find the area if one side is 0.12 inch long.

_____

Unit 19   SCIENTIFIC NOTATION

BASIC PRINCIPLES OF SCIENTIFIC NOTATION

Scientific notation is used in almost all scientific calculations. It was first used for making calculations with a slide rule. A slide rule is a tool that can perform mathematical operations such as multiplication and division, find square roots, find logs of numbers, and find sines, cosines, and tangents of angles. When finding a number on a slide rule, only the actual digits are used, not decimal points or zeros (unless they come between two other digits such as 102). To a slide rule, the numbers 0.000012, 0.0012, 0.12, 1.2, 12, 120, 12000, and 12000000 are all the same number: 12.

Since the slide rule recognizes only the basic digits of any number, imagine the problem of determining where to place a decimal point in an answer. As long as only simple calculations are done, there is no problem in determining where the decimal point should be placed.

Example: Multiply 12×20

It is obvious where the decimal point should be placed in this problem.

$$240.00$$

Example: Now assume that the following numbers are to be multiplied together:

$$0.000041 \times 380,000 \times 0.19 \times 720 \times 0.0032$$

In this problem, it is not obvious where the decimal point should be placed in the answer. Scientific notation can be used to simplify the numbers so that an estimated answer can be obtained. Scientific notation is used to change any number to a simple whole number by dividing or multiplying the number by a power of 10. Any number can be multiplied by 10 by moving the decimal point one place to the right. Any number can be divided by 10 by moving the decimal point one place to the left.

Example: The number 0.000041 can be changed to a whole number of 4.1 by multiplying it by 10 five times. Therefore, if the number 4.1 is divided by 10 five times, it will be the same as the original number, 0.000041. In this problem, the number 0.000041 will be changed to 4.1 by multiplying it by 10 five times. The new number is: 4.1 times 10 to the negative fifth.

$$4.1 \times 10^{-5}$$

Since superscripts and subscripts are often hard to produce in printed material, it is a common practice to use the letter E to indicate the exponent in scientific notation. The previous notation can also be written:

$$4.1E\text{-}05$$

The number 380,000 can be reduced to a simple whole number of 3.8 by dividing it by 10 five times. The number 3.8, therefore, must be notated to indicate that the original number is actually 3.8 multiplied by 10 five times:

$$3.8 \times 10^{+5}$$

or

$$3.8E\ 05$$

The other numbers in the problem can be changed to simple whole numbers using scientific notation:

$$0.19 \text{ becomes } 1.9E\text{-}01$$

$$720 \text{ becomes } 7.2E\ 02$$

$$0.0032 \text{ becomes } 3.2E\text{-}03$$

Now that the numbers have been simplified using scientific notation, an estimate can be obtained by multiplying these simple numbers together and adding the exponents: 4.1 is about 4; 3.8 is about 4; 1.9 is about 2; 7.2 is about 7; and 3.2 is about 3. Therefore:

$$4 \times 4 \times 2 \times 7 \times 3 = 672$$

When the exponents are added

$$(E\text{-}05) + (E\ 05) + (E\text{-}01) + (E\ 02) + (E\text{-}03) = E\text{-}02$$

the estimated answer becomes 672E-02. When the calculation is completed, the actual answer becomes:

$$682.03E\text{-}02$$

or

$$6.8203$$

USING SCIENTIFIC NOTATION WITH CALCULATORS

In the early 1970s, scientific calculators, often referred to as *slide rule calculators,* became commonplace. Most of these calculators have the ability to display from eight to ten digits depending on the manufacturer. Scientific calculations, however, often involve numbers that contain more than eight or ten digits. To overcome the limitation of an eight- or ten-digit display, slide rule calculators depend on scientific notation. When a number becomes too large for the calculator to display, scientific notation is used automatically. Imagine, for example, that it became necessary to display the distance (in kilometers) that light travels in one year (approximately 9,460,800,000,000 kilometers). This number contains thirteen digits. The calculator would display this number as shown:

$$9.4608 \ 12$$

The number 12 shown to the right of 9.4608 is the scientific notation exponent. This number could be written $9.4608E12$, indicating that the decimal point should be moved to the right twelve places.

If a minus sign should appear ahead of the scientific notation exponent, it indicates that the decimal point should be moved to the left. The number on the following display contains a negative scientific notation exponent.

$$7.5698 \ -06$$

This number could be written as:

$$0.0000075698$$

Entering Numbers in Scientific Notation

Slide rule calculators also have the ability to enter numbers in scientific notation. To do this, the exponent key must be used. There are two general ways in which manufacturers mark the exponent key. Some are marked **EXP** and others are marked **EE.**

Example: Assume that the number of $549E \ 08$ was to be entered. The following key strokes would be used:

$$\boxed{5} \ \boxed{4} \ \boxed{9} \ \boxed{EE} \ \boxed{8}$$

The calculator would display the following:

$$549 \ 08$$

If a number with a negative exponent is to be entered, the change sign (+/−) key should be used. Assume that the number 1.276E-4 is to be entered. The following key strokes would be used:

$$\boxed{1}\ \boxed{\bullet}\ \boxed{2}\ \boxed{7}\ \boxed{6}\ \boxed{\text{EXP}}\ \boxed{4}\ \boxed{+/-}$$

The display would show the following:

$$1.276\ -04$$

Setting the Display

Some calculators permit the answer to be displayed in any of three different ways. One of these ways is with *floating decimals* (FD). When the calculator is set for this mode of operation, the answers will be displayed with the decimal point appearing in the normal position. The only exception to this is if the number to be displayed is too large. In this case, the calculator will automatically display the number in scientific notation.

In the *scientific* mode (Sci), the calculator will display all entries and answers in scientific notation. When set in the *engineering* mode (Eng), the calculator will display all entries and answers in scientific notation, but only in steps of three. When displayed in steps of three, the notation corresponds to standard engineering notation units such as kilo, mega, giga, milli, micro, and so on. For example, assume a calculator is set in the scientific mode and displays the number shown:

$$5.69836\ 05$$

Now assume that the calculator is reset to the engineering mode. The number would now be displayed as shown:

$$569.836\ 03$$

The number could now be read as "569.836 kilo" because kilo means one thousand, or 1E03.

PRACTICAL PROBLEMS

Express each of the following numbers in scientific notation.

1. $100 \times 10^{15} =$ _____

2. $100 \times 10^{-15} =$ _____

3. $63.25 \times 10^{-4} =$ _____

4. $810.36 \times 10^{7} =$ _____

5. $0.000\ 000\ 000\ 003 \times 10^6 =$ _____

6. $725 \times 10^{19} =$ _____

7. $0.36 \times 10^4 =$ _____

8. $482{,}600{,}000 \times 10^{-5} =$ _____

9. $0.005 \times 10^{12} =$ _____

10. $11.56 \times 10^{-9} =$ _____

Note: Use this schematic to express the circuit values in problems 11–20 in scientific notation.

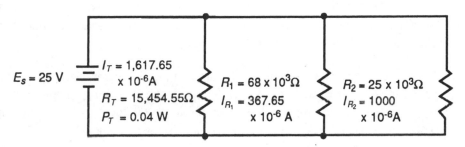

$R_3 = 25 \times 10^3 \Omega$
$I_{R_3} = 250 \times 10^{-6} A$

$E_s = 25\ V$

$I_T = 1{,}617.65 \times 10^{-6} A$
$R_T = 15{,}454.55 \Omega$
$P_T = 0.04\ W$

$R_1 = 68 \times 10^3 \Omega$
$I_{R_1} = 367.65 \times 10^{-6} A$

$R_2 = 25 \times 10^3 \Omega$
$I_{R_2} = 1000 \times 10^{-6} A$

11. $E_s =$ _____

12. $R_1 =$ _____

13. $R_2 =$ _____

14. $R_3 =$ _____

15. $R_T -$ _____

16. $I_T =$ _____

17. $I_{R_1} =$ _____

18. $I_{R_2} =$ _____

19. $I_{R_3} =$ _____

20. $P_T =$ _____

Unit 20 EQUIVALENT ELECTRONIC UNITS

BASIC PRINCIPLES OF ELECTRONIC UNITS

Scientific measurements use units that are based on the *metric system*. These units, referred to as *engineering units* or *engineering notation,* are graduated in steps of 1,000 instead of steps of 10. When using engineering notation, the first unit above the base unit is *kilo*, which means 1,000, and is symbolized by the letter *k*. 5 kilovolts, for example, means 5,000 volts. The next engineering notation unit is *mega,* or million. Mega is 1,000 times greater than kilo. Mega is symbolized by the letter *M.* The next engineering unit above mega is *giga,* which is 1,000 mega, or billion. Giga is symbolized by the letter *G.* Microwave frequencies are often measured in gigahertz (GHz). The average microwave oven operates at a frequency of about 2.45 GHz, which is 2,450,000,000 hertz (Hz). The engineering unit above giga is *terra,* which is 1,000 giga, or trillion. Terra is symbolized by the letter *T.*

The first engineering unit below the base unit is *milli* or one thousandth. Milli is symbolized by the letter *m.* The engineering unit below milli is *micro,* which is 1,000 times smaller than milli, or one millionth. Micro is symbolized by the Greek letter μ. The next unit is 1,000 times smaller than micro and is called *nano.* Nano, meaning one billionth, is symbolized by the letter *n.* The engineering unit below nano is *pico,* which is 1,000 times less than nano, or one trillionth. Pico is symbolized by the letter *p.* Since pico is actually 1,000,000 times less than micro, this measurement is sometimes referred to as micro micro, and is symbolized by $\mu\mu$.

Terra	1,000,000,000,000	T	10^{12}
Giga	1,000,000,000	G	10^{9}
Mega	1,000,000	M	10^{6}
Kilo	1,000	k	10^{3}
Unit value	1		
milli	0.001	m	10^{-3}
micro	0.000 001	μ	10^{-6}
nano	0.000 000 001	n	10^{-9}
pico	0.000 000 000 001	p	10^{-12}

PRACTICAL PROBLEMS

1. A radio station broadcasts on 96 megahertz (MHz). Express this frequency in hertz (Hz).

    _____

2. The ammeter in this circuit indicates the number of amperes of current. How many milliamperes (mA) are drawn?

    _____

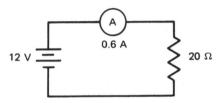

12 V · 0.6 A · A · 20 Ω

3. A resistor is marked 0.47 megohm (MΩ). How many ohms (Ω) is this resistor?

    _____

4. A high-range voltmeter reads 24,000 volts (V). Express this voltage in kilovolts (kV).

    _____

5. The capacitor in this diagram is marked in microfarads (μF). Express the capacitance in picofarads (pF).

    _____

0.002

6. UHF television channel 34 operates at 590 megahertz (MHz). Express the frequency in hertz (Hz).

    _____

7. The current from the transistor in the circuit is called base current (I_b). What is I_b in milliamperes?

    _____

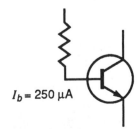

$I_b = 250$ μA

8. A lethal amount of current can be as low as 150 milliamperes. Express this current value in amperes.   _____

9. The meter scale indicates voltage from 0 volt to 1 volt. What is this voltage reading in millivolts?   _____

10. A fuse will blow and open a circuit if current is larger than the fuse rating. What is the rating in amperes of the fuse in the diagram?   _____

11. The output of an oscillator is 47,520 hertz (Hz). Express the output frequency in kilohertz (kHz).   _____

12. A 1-megohm (MΩ) potentiometer is set to 0.8 megohm. Express this resistance in ohms.   _____

13. The inductor in the diagram is replaced by an inductor marked in microhenries (µH). How many microhenries are marked on the inductor?   _____

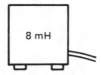

14. When light strikes a solar cell, 358 millivolts are produced. Express this voltage in volts.   _____

15. Sound travels 1,100 feet per second. In ¹⁄₂₀ second, it travels 55 feet. Express this time in microseconds (µs).   _____

16. The diode in the diagram has a voltage drop across it and a current through it.

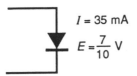

$I = 35$ mA

$E = \frac{7}{10}$ V

   a. Express the voltage in millivolts.   a. _____

   b. Express the current in amperes.   b. _____

17. A resistor has a color code of green, blue, and red. This stands for 5,600 ohms (Ω). How many kilohms (kΩ) is this?   _____

18. A radio station transmits 50,000 watts (W) of power. Express this power in megawatts (MW).   _____

19. A radar transmitter is set to a frequency of 9,400 megahertz (MHz). Express this frequency in kilohertz (kHz).   _____

20. The display of a digital voltmeter indicates 0.072 8 volt. If the meter is set to display millivolts, what is the reading?   _____

Unit 21   OPERATIONS WITH ELECTRONIC UNITS

BASIC PRINCIPLES OF USING ELECTRONIC UNITS

When calculating electrical values, it is not uncommon to encounter different units of measure in the same problem. When this is the case, it is necessary to change all values to the same unit of measure. Assume, for example, that two resistors are connected in series and that it is necessary to find their total resistance. Also assume that one resistor has a value of 510 ohms and the other has a value of 2.2 kilohms. Notice that the value of one resistor is given in ohms, and the other is given in kilohms. Before the total resistance can be found, it will be necessary to change both values into the same unit of measure. This can be done in either of two ways. One way is to change 2.2 kilohms into the base unit by multiplying 2.2 by 1,000. This will change 2.2 into 2,200.

Example: 2.2 kilohms $\times$ 1,000 ohms/kilohm = 2,200 ohms

510 ohms + 2,200 ohms = 2,710 ohms

The second way is to change 510 ohms into kilohms. This can be done by dividing 510 by 1,000.

510 ohms $\div$ 1,000 ohms/kilohm = 0.510 kilohms

2.2 kilohms + 0.510 kilohm = 2.71 kilohms

Although the two answers look different, they are both the same. 2.71 kilohms is the same as 2,710 ohms.

PRACTICAL PROBLEMS

1. The values of resistors in series are added to form total resistance (R_t). Find R_t in kilohms in this circuit.   _____

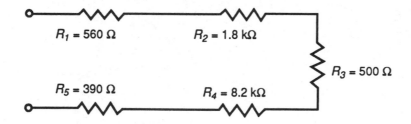

$R_1 = 560\ \Omega$ $R_2 = 1.8\ k\Omega$

$R_3 = 500\ \Omega$

$R_5 = 390\ \Omega$ $R_4 = 8.2\ k\Omega$

2. An electronic mixer can subtract two frequencies. The inputs are 0.8 megahertz and 345 kilohertz. Find the difference, in kilohertz, between the frequencies.

_____

3. The output current of an amplifier is the input current (I_b) multiplied by the circuit gain. Use this circuit to find the output current (I_c) in milliamperes if the gain (β) is 65.

_____

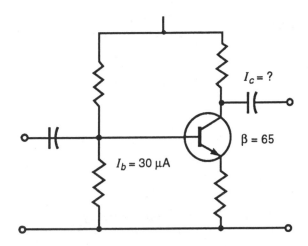

4. The quality (Q) of a tuned circuit is inductive reactance (X_L) divided by resistance (R). Find the Q of this circuit.

    Note: $Q = X_L \div R$

_____

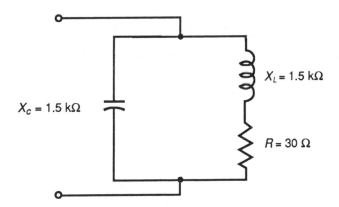

5. The values of inductors connected in series are added to form total inductance (L_t). Find L_t in henries (H) in this circuit.   _____

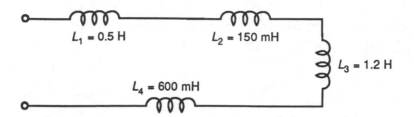

$L_1 = 0.5$ H $L_2 = 150$ mH

$L_3 = 1.2$ H

$L_4 = 600$ mH

6. In transistors, the emitter current (I_e) equals base current (I_b) plus collector current (I_c). Find I_b in microamperes if $I_c = 3.76$ milliamperes and $I_e = 3.82$.   _____

    Note: $I_b = I_e - I_c$

7. The peak voltage (E_p) of an AC signal is 1.5 volts. Find the rms voltage in millivolts.   _____

    Note: $E_{rms} = E_p \times 0.707$

8. Find the period (T) of a signal in microseconds if the frequency (f) is 50 kilohertz.   _____

    Note: $T = \dfrac{1}{f}$

9. Branch currents in a parallel circuit are added to form total circuit current (I_t). Use the schematic to find I_t in milliamperes.   _____

$I_t = ?$

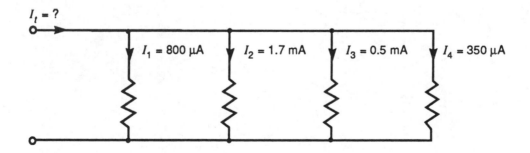

$I_1 = 800$ μA $I_2 = 1.7$ mA $I_3 = 0.5$ mA $I_4 = 350$ μA

10. A 2.5 megohm potentiometer is set to 780 kilohm in one portion of the pot. What is the amount of resistance in kilohms in the other portion of the pot?   _____

11. An oscillator generates AC signals at an output frequency of 20,000 cycles per second, or hertz (Hz). What would be the frequency in kilohertz (kHz) if the signal was measured over a period of one minute?   _____

12. The power gain of an amplifier is the output power divided by the input power. Find the gain if the input is 4.5 milliwatts and output is 0.117 watt.   _____

13. A digital voltmeter is used to measure voltages in the series circuit shown. The sum of the voltage values is the source voltage (E_s). Find E_s in volts for this circuit.   _____

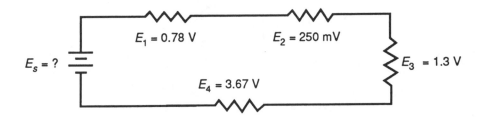

$E_1 = 0.78$ V $E_2 = 250$ mV

$E_s = ?$

$E_3 = 1.3$ V

$E_4 = 3.67$ V

14. A cable television company must run 0.82 kilometer of wire to connect a farmhouse. The workers have 650 meters of wire on the truck. How many additional meters are needed?   _____

15. The peak-to-peak voltage (E_{pp}) of an AC signal is 2.828 times the rms voltage (E_{rms}). Find the peak-to-peak voltage in volts of a 100-millivolt rms signal.   _____

16. Find the frequency (f) in kilohertz of a signal with a period (T) of 25 microseconds.   _____

    Note: $f = \dfrac{1}{T}$

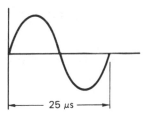

25 μs

17. An electronic mixer can add the frequencies put into it. The inputs are 4.8 megahertz and 650 kilohertz. Find the output in megahertz (MHz).   _____

18. Voltage drops in a series circuit are added to equal source voltage. Find the value of E_2 in volts in the circuit.

Note: $E_2 = E_s - E_1 - E_3 - E_4$

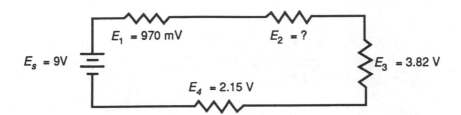

19. Operational amplifiers have a very high voltage gain. The output in this circuit is 4,500 times the input. Find the output voltage in volts in this op-amp circuit.

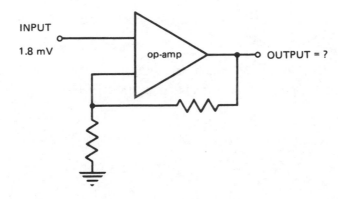

20. The voltage gain of an operational amplifier is found by dividing output voltage by input voltage. The output value is 4.785 volts and the input value is 580 microvolts. Compute the gain.

21. The formula for determining the amount of current needed to melt copper wire is:

$$I_{melt} = K\sqrt{D^3}$$

where K = 10244, a constant for copper
D = the diameter of the wire in inches

How many amperes are required to melt a piece of number 14 AWG copper wire if the circular mil area is 4,107 circular mils (CM)? (Hint: 1 mil = 0.001 inch, $D^2 = (1mil)^2 = 1CM$)

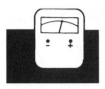

Unit 22   ROOTS

BASIC PRINCIPLES OF ROOTS

The *root* of a number is the opposite or inverse of its power. The *square root* of a number is determined by finding which number used as a factor twice will equal the given number. The square root of 36 is 6 because 6 squared equals 36.

$$(6 \times 6) = 36$$

The process is similar for other roots also. The *cube root* of 27 is 3 because 3 cubed equals 27.

$$(3 \times 3 \times 3) = 27$$

When the root of a number is to be found, the number is placed under a *radical sign* ($\sqrt{}$). If the square root of a number is to be found, the problem is written like this:

$$\sqrt{36} = 6$$

If some root other than a square root is to be found, a small number is placed outside the radical sign. This number indicates what root is to be found.

$$\sqrt[3]{27} = 3$$

Finding Roots with a Calculator

Most scientific calculators can be used to find roots of numbers. Almost all calculators contain a square root key ($\sqrt{x}$). To find the square root of a number, simply input the number and press the $\sqrt{x}$ key.

Example: Find the square root of 441.

$$\boxed{4}\ \boxed{4}\ \boxed{1}\ \boxed{\sqrt{x}}$$

The answer is 21.

It is sometimes necessary to find other roots of numbers. Assume, for example, that you need to find the fourth root of 50,625 ($\sqrt[4]{50,625}$). How you do this will depend on the type of calculator you use. Although most scientific calculators have the ability to find any root of any number, not all do it the same way. Some calculators contain an *nth* root key:

$$\boxed{\sqrt[x]{y}}$$

To find the fourth root of 50,625 using a calculator with an nth root key, input 50625, press the nth root key, then press 4, and then press =.

Example: 5 0 6 2 5 $\sqrt[x]{y}$ 4 =

The answer is 15.

If the calculator does not contain an nth root key, it will almost certainly contain an nth power key:

y^x

There are two methods of using the nth power key to find an nth root. If the calculator contains an invert key (INV), it can be used to change the nth power function into an nth root function. To find the fourth root of 50,625 using a calculator of this type, follow the procedure shown:

Example: 5 0 6 2 5 INV y^x 4 =

For calculators that do not contain the invert key, the nth power key can be used to find an nth root by using the reciprocal key ($\frac{1}{x}$). Any number raised to the reciprocal of any power will be the same as finding the root. Raising 50,625 to the one-fourth power is the same as finding the fourth root of 50,625. That is,

$$\sqrt[4]{50{,}625} = 50{,}625^{\frac{1}{4}} = 50{,}625^{0.25}$$

Example: 5 0 6 2 5 y^x 4 $\frac{1}{x}$ =

The answer is 225.

PRACTICAL PROBLEMS

1. $\sqrt[3]{27}$ _____

2. $\sqrt[4]{16}$ _____

3. $\sqrt[3]{343}$ _____

4. $\sqrt[5]{1}$ _____

5. $\sqrt[4]{81}$ _____

6. $\sqrt[6]{64}$ _____

7. $\sqrt[3]{64}$ _____

8. $\sqrt[3]{125}$ _____

9. $\sqrt[4]{1296}$ _____

10. $\sqrt[3]{216}$ _____

11. A number of transformers are mounted on a square plate with an area (A) of 784 square centimeters (cm²). Find the length of one side.   _____

Note: $s = \sqrt{A}$

12. The tank in the diagram is used to store etchant for circuit boards. Find the length of one side of the cubic tank.   _____

Note: $s = \sqrt[3]{V}$

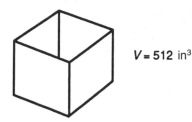

$V = 512$ in³

13. One form of Ohm's law states voltage (E) equals the square root of the product of the power and resistance. Find E in volts if power times resistance is 144.   _____

Note: $E = \sqrt{PR}$

14. Find the length of one side of the integrated circuit chip in the diagram.   _____

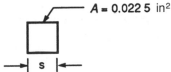

$A = 0.022\ 5$ in²

s

15. Fred and Jim are working an electronics problem. The question asks for the square root of 73.96. What is the answer?   _____

Note: Use this information for problems 16–18. The cross-sectional area (A) of a piece of wire is given in circular mils (CM). The wire diameter (d), in mils, is the square root of the circular mil area; that is, $d = \sqrt{A}$.

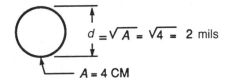

$d = \sqrt{A} = \sqrt{4} = 2$ mils

$A = 4$ CM

16. A piece of 20-gauge wire has a cross-sectional area of 1,024 circular mils (CM). Find the diameter of the wire in mils.   _____

17. Compute the diameter in mils for the wire in this diagram.   _____

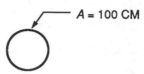

$A = 100$ CM

18. Three 25-gauge wires are run next to each other. The cross-sectional area of each is approximately 324 circular mils. Use this diagram to find the length X in mils.   _____

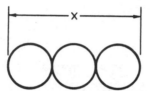

19. The area (*A*) of a square circuit board is 22⁹⁄₁₆ square inches. Find the length of one side.   _____

   Note: $s = \sqrt{A}$

20. The relay in this diagram is made in the shape of a cube. Find the length of the side (*s*) in inches if the volume (*V*) is 3⅜ cubic inches.   _____

   Note: $s = \sqrt[3]{V}$

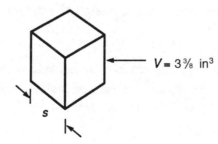

$V = 3\frac{3}{8}$ in^3

s

Formulas, Ohm's Law, and Power Law

Unit 23   EQUATIONS

BASIC PRINCIPLES OF REPRESENTATION IN FORMULAS

A *formula* is a mathematical statement of equality. Just as words are written using symbols, formulas are also written using symbols. In the following problems, written statements will be expressed as mathematical formulas.

When writing a formula, it is helpful to know what some of the statements mean and how some of the symbols are used to indicate different operations. The word *sum* is used to indicate addition, and a plus sign (+) is used to represent addition.

Subtraction is generally indicated by statements such as *the difference of.* The minus sign (−) is generally used to represent subtraction.

The word *product* is used to indicate multiplication. There are several ways to indicate that two quantities are to be multiplied together. One way is to simply write two variables (alphabetical characters used to represent the quantities) together with no sign between them. *IR* means to multiply *I* by *R*. Another method is to use parentheses around the values to be multiplied together. (*I*) (*R*) means to multiply *I* and *R* together. Symbols are also used to represent multiplication. The multiplication sign (×) is often used between numerals but is seldom used with variables. The multiplication dot (•) is often used between two variables to represent multiplication. Another symbol used to represent multiplication, the asterisk (*), has become popular because of computers.

The words *ratio* and *proportional* are represented by division. Division is generally indicated in a formula by writing the numbers as a fraction. If the letter *E* is to be divided by *R,* it would be written as

$$\frac{E}{R}$$

In the formula:

$$I = \frac{E}{R}$$

I is *directly proportional* to *E* and *inversely proportional* to *R.*

Another term often used in mathematics is *reciprocal.* The reciprocal of a number means 1 divided by the number. The reciprocal of 4 is

$$\frac{1}{4}$$

Equations

Mathematical formulas are generally referred to as *equations,* which means *statements of equality.* To better understand what an equation is, think of a balance scale with the equal sign being the center or balance point. The scales can be balanced only when both sides have the same weight or value. In order for an equation to be valid, both sides must remain in balance. Basically, anything can be done to one side of the equation as long as the same thing is done to the other side. If five ounces is added to both sides of the scales, they will remain in balance. This five ounces of weight could be added in the form of a five-ounce weight on one side and a two-ounce and a three-ounce weight on the other side. Values can be added, subtracted, multiplied, or divided on one side of the scale as long as the same is done on the other side.

Addition and subtraction of the same value on both sides of an equation maintains equivalent expressions.

In algebra, a letter used to represent an unknown quantity is called a *variable.* Variables do not have a fixed value. The value of the variable can change in accord with other factors in the equation. Variables are used to produce formulas for finding unknown values when other values are known. A very common formula in Ohm's law is $E = IR$. This formula states that voltage (E) is equal to current (I) multiplied by resistance (R). The formula permits different numerical values to be substituted for the letters. For example, assume that a 100-ohm resister has 50 milliamperes of current flowing through it. What is the voltage drop across the resistor? To find the answer, substitute the known numerical values for the corresponding variables in the equation:

$$E = I \bullet R$$
$$E = 0.050 \bullet 100$$
$$E = 5 \text{ V}$$

Examples: Solve these equations.

a. Solve the equation $4 + x = 12$ for x.

$$4 + x = 12$$ (original equation)

$$4 + x - 4 = 12 - 4$$ (subtract 4 from *both* sides)

$$x = 8$$ (solution)

b. Solve the equation $z - 23 = 36$ for z.

$$z - 23 = 36$$ (original equation)

$$z - 23 + 23 = 36 + 23$$ (add 23 to *both* sides)

$$z = 59$$ (solution)

Multiplication or division of the same value (except zero) on both sides of an equation maintains equivalent expressions.

Examples: Solve these equations.

a. Solve $\dfrac{y}{18} = 5$ for y.

$$\dfrac{y}{18} = 5$$ (original equation)

$$\dfrac{y}{18} \times 18 = 5 \times 18$$ (multiply *both* sides by 18 so that the 18s on the left side "cancel," leaving y)

$$y = 90$$ (solution)

b. Solve $9.5a = 21.85$ for a.

$$9.5a = 21.85$$ (original equation)

$$\dfrac{9.5a}{9.5} = \dfrac{21.85}{9.5}$$ (divide *both* sides by 9.5 so the 9.5s will "cancel," leaving the solution for a)

$$a = 2.3$$

There will be many equations involving fractions. A common method employed in solving these equations is known as *cross-multiplication*. This is an application of the previous rule. Follow this example carefully.

Example: Solve for *b* in this fraction.

$$\frac{24}{b} = \frac{60}{10}$$ (original equation)

$$\frac{24}{b} \diagdown \frac{60}{10}$$ (numerators and denominators are cross-multiplied)

$$24 \times 10 = 60 \times b$$ (products of cross-multiplication)

$$240 = 60b$$

$$\frac{240}{60} = b$$ (divide *both* sides by 60)

$$b = 4$$

Rearrangement in Formulas

It is sometimes necessary to rearrange formulas so that different quantities can be found. This is done by isolating the quantity to be found on one side of the equal sign. Although it makes no difference on which side of the equal sign the unknown quantity is located, it is common practice to place the quantity to be found on the left hand side of the equal sign. When rearranging formulas, remember two basic principles.

1. Anything can be done to one side of an equation as long as the same thing is done to the other side.

2. When moving factors from one side of the equation to the other, perform the inverse of the operation indicated.

Example: $X = 2JLC$. Solve for *L*.

L must be on one side of the equal sign by itself. This can be done by removing all the other factors on the right side of the equation. The formula states that *X* is equal to 2 times *J* times *L* times *C*. Multiplication is the operation indicated. To remove all the factors except *L*, perform division, which is the inverse operation of multiplication.

$$\frac{X}{2JC} = \frac{2JLC}{2JC}$$

Notice that both sides of the equation are divided by 2*JC*, which will leave *L*.

$$\frac{X}{2JC} = \frac{\overset{L}{\cancel{2JLC}}}{\underset{1}{\cancel{2JC}}}$$

The equation is now

$$\frac{X}{2JC} = \frac{L}{1}$$

The formula can now be written as

$$L = \frac{X}{2JC}$$

PRACTICAL PROBLEMS

Solve each of the following equations for the unknown value.

1. $3x = 12$ $x =$ _____

2. $8y + 6 = 3$ $y =$ _____

3. $\dfrac{x}{5} = 25$ $x =$ _____

4. $48 + m = 61$ $m =$ _____

5. $12 + p + 3p = 67.2$ $p =$ _____

6. $\dfrac{x + 3}{7} = 9$ $x =$ _____

7. $5y - 19 = 154$ $y =$ _____

8. $0.9 + z = 4.7$ $z =$ _____

9. $19 + 5a = 35.25$ $a =$ _____

10. $3b + 14 = 5b$ $b =$ _____

11. $\dfrac{8 + c}{2.29} = 4$ $c =$ _____

12. $4.5m - 3 = 2m - 0.5$ $m =$ _____

13. $b - 12 = 12$ $b = $ _____

14. $4y + 2 = 3.2$ $y = $ _____

15. $\dfrac{4y + 2}{6} = 21$ $y = $ _____

16. $7.3m - 16 = m + 0.38$ $m = $ _____

17. $a + 4 = 3a - 81$ $a = $ _____

18. $1 + \dfrac{x}{5} = 8$ $x = $ _____

19. $\dfrac{x}{5} - 1 = 6$ $x = $ _____

20. $\dfrac{3x + 9}{4} = 3.6$ $x = $ _____

21. $5 + 12z = 20z + 1$ $z = $ _____

22. $146 = \dfrac{36y + 2}{4} = 236.5$ $y = $ _____

23. $z + 4 = 4$ $z = $ _____

24. $18n + 2n = 6n + 11.2$ $n = $ _____

25. $\dfrac{28x - 4}{8} + 36 = 63.22$ $x = $ _____

Solve for the indicated variable in each of the following electronics formulas.

26. $E = \dfrac{P}{I}$, solve for I _____

27. $E = IR$, solve for R _____

28. $R = \dfrac{KL}{A}$, solve for L _____

29. $G = \dfrac{1}{R}$, solve for R _____

30. $R_s = \dfrac{I_m \times R_m}{I_s}$, solve for R_m _____

31. $X_L = 2\pi fL$, solve for f

————————

32. $X_c = \dfrac{1}{2\pi fC}$, solve for C

————————

33. $L_t = L_1 + L_2 + 2L_m$, solve for L_m

————————

34. $I_B = I_{BQ} + \sqrt{2I_s}$, solve for I_s

————————

35. $T_{max} = 5RC$, solve for R

————————

36. $C = \dfrac{Q}{E}$, solve for E

————————

37. $Q = \dfrac{X_L}{R}$, solve for X_L

————————

38. $E_{ms} = 0.707 \times \dfrac{E_{pp}}{2}$, solve for E_{pp}

————————

39. $\lambda = \dfrac{300,000,000}{f}$, solve for f

————————

40. $L_t = L_1 + L_2 - 2L_m$, solve for L_m

————————

41. $E_s = E_1 + E_2 + E_3$, solve for E_2

————————

42. $R_1 = \dfrac{I_2 R_2}{I_2}$, solve for R

————————

43. $°F = (°C \times \dfrac{9}{5}) + 32°$, solve for $°C$

————————

44. $\dfrac{R_1}{R_x} = \dfrac{R_2}{R_3}$, solve for R_x

————————

45. $\dfrac{1}{R_T} = \dfrac{1}{R_1} + \dfrac{1}{R_2}$, solve for $\dfrac{1}{R_1}$

————————

46. $R = \dfrac{E^2}{P}$, solve for P

————————

47. $y = mx + b$, solve for x

————————

48. $y = mx + b$, solve for b

————————

49. $I_2 = \dfrac{I_1 R_1}{R_2}$, solve for I_1

————————

50. $C = (8.85 \times 10^{-12})\,(K)\,(\dfrac{A}{d})$, solve for d

————————

Unit 24 FORMULAS

Review and apply the principles presented in Section 4 and used in Unit 23 to these problems. Study the *Units and Tables of Electronic Measurement* in Section II of the Appendix.

PRACTICAL PROBLEMS

Compute the indicated values. If necessary, round the answers to the nearer thousandth.

1. Find the resistance of 150 feet of 22-gauge copper wire. The diameter (d) of 22-gauge wire is 25.3 mils. _____

   Note: $R = \dfrac{kL}{A}$ where R = resistance in ohms
   k = resistivity of copper ($10.7\ \dfrac{\Omega CM}{ft}$)
   L = length in feet
   A = area in circular mils (CM) = d^2

2. Find the total capacitance (C_t) in this circuit. _____

   Note: $C_t = \dfrac{C_1 C_2}{C_1 + C_2}$

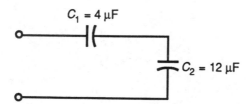

$C_1 = 4\ \mu F$

$C_2 = 12\ \mu F$

3. Conductance (G) is the opposite of resistance and is measured in siemens. Find the conductance in siemens of a 200-ohm resistor. _____

   Note: $G = \dfrac{1}{R}$

4. Find total inductance (L_t) in this circuit assuming there is no mutual coupling. (L_1 and L_2 must be in the same units.) _____

    Note: $L_t = \dfrac{L_1 L_2}{L_1 + L_2}$

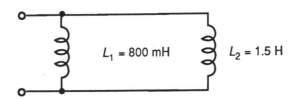

$L_1 = 800$ mH $L_2 = 1.5$ H

5. Find one time constant (τ) in milliseconds in a series *RL* circuit. The inductor (*L*) is 0.5-henry and the resistor (*R*) is 100 ohms. (Henries divided by ohms equals seconds.) _____

    Note: $\tau = \dfrac{L}{R}$

6. It takes 5 time constants for the *RC* circuit in the diagram to fully charge. Find the time in milliseconds to fully charge the circuit. (Kilohms times microfarads equals milliseconds.) _____

    Note: $T_{max} = 5\tau = 5RC$ where T_{max} = 5 time constants in seconds

    $\quad R$ = resistance in ohms

    $\quad C$ = capacitance in farads

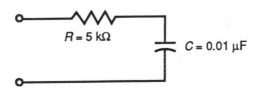

$R = 5$ kΩ $C = 0.01$ μF

7. Capacitance (*C*) can be found by dividing the stored charge (*q*) by the voltage (*E*). Find capacitance in farads if 0.3 coulomb of charge is stored across 12 volts. _____

    Note: $C = \dfrac{q}{E}$

8. Compute total capacitance (C_t) in this series-parallel circuit. _____

    Note: $C_t = \dfrac{C_1\,(C_2 + C_3)}{C_1 + C_2 + C_3}$

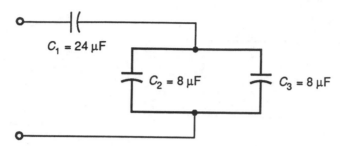

9. The quality (Q) of a tuned circuit is inductive reactance (X_L) divided by resistance (R). Find Q of a circuit with $X_L = 280$ ohms and $R = 17\frac{1}{2}$ ohms. _____

    Note: $Q = \dfrac{X_L}{R}$

10. Shunt resistors are used in constructing ammeters. Find the value of the shunt resistor in this circuit. _____

    Note: $R_s = \dfrac{I_m R_m}{I_s}$ where R_s = shunt resistance in ohms
    I_m = full scale meter current in amperes
    R_m = meter resistance in ohms
    I_s = shunt current in amperes

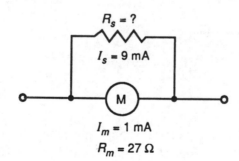

11. What is the rms voltage (E_{rms}) of an AC signal with a peak-to-peak voltage (E_{pp}) of 340 volts? _____

    Note: $E_{rms} = 0.353\,5E_{pp}$

12. The peak-to-peak voltage on an oscilloscope is 84 volts. Calculate the rms voltage. _____

    Note: $E_{rms} = 0.353\,5E_{pp}$

13. Find the E_{pp} of a voltage with an rms value of 12 volts.   _____

   Note: $E_{pp} = \dfrac{E_{rms}}{0.353\ 5}$

14. Find the wavelength (λ) in meters of an FM radio signal with a frequency (f) of 96 megahertz.   _____

   Note: $\lambda = \dfrac{300{,}000{,}000}{f}$

15. Calculate the total inductance (L_t) in megahenries in this circuit assuming there is no mutual coupling.   _____

   Note: $L_t = L_1 + \dfrac{L_2 L_3}{L_2 + L_3}$

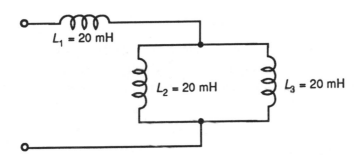

$L_1 = 20$ mH

$L_2 = 20$ mH $L_3 = 20$ mH

16. What is the mutual inductance (L_m) in henries (H) between two inductors with $L_1 = 3$ H and $L_2 = 0.48$ H?   _____

   Note: $L_m = 0.65\sqrt{L_1 L_2}$

17. Compute the bandwidth (BW) in hertz (Hz) of a circuit with a resonant frequency (f_o) of 10,000 hertz. The quality (Q) of the circuit is 12.5.   _____

   Note: $BW = \dfrac{f_o}{Q}$

18. What is the base voltage (E_b) in this transistor circuit?   _____

Note: $E_b = E_{cc} \times \dfrac{R_2}{R_1 + R_2}$

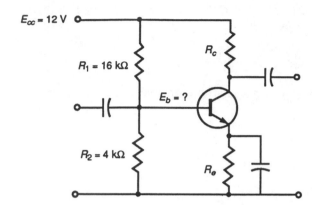

19. Beta (β) is the current gain in a common emitter circuit. It is the change in
     collector current (I_c) divided by the change in base current (I_b). Find beta
     when I_b changes from 25 µA to 75 µA and I_c changes from 2 mA to 6 mA.   _____

Note: $\beta = \dfrac{\text{change in } I_c}{\text{change in } I_b}$

20. What is the value of the voltage across R_L?   _____

Note: $E_{R_L} = \dfrac{R_L}{R_1 + R_L} \times E_s$

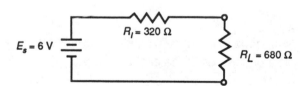

21. Inductors of 75 millihenries and 120 millihenries are connected in series-
     aiding. They have a mutual inductance (L_m) of 42 millihenries. Find total
     inductance (L_t).   _____

Note: $L_t = L_1 + L_2 + 2L_m$

22. Inductors of 320 microhenries and 265 microhenries are connected in series-
     opposing. Find total inductance (L_t) if the mutual inductance (L_m) is 82
     microhenries.   _____

Note: $L_t = L_1 + L_2 - 2L_m$

23.   Find the values for the currents indicated.

Note: $I_1 = I_t \times \dfrac{R_2}{R_1 + R_2}$

$I_2 = I_t - I_1$

a. I_1

b. I_2

a. _____

b. _____

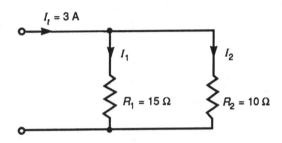

$I_t = 3\ \text{A}$

I_1 I_2

$R_1 = 15\ \Omega$ $R_2 = 10\ \Omega$

24.   Compute these voltages and check E_s in this circuit.

_____

Note: $E_1 = E_s \times \dfrac{R_1}{R_1 + R_2 + R_3}$

$E_2 = E_s \times \dfrac{R_2}{R_1 + R_2 + R_3}$

$E_3 = E_s \times \dfrac{R_3}{R_1 + R_2 + R_3}$

$E_s = E_1 + E_2 + E_3$

a. E_1

b. E_2

c. E_3

a. _____

b. _____

c. _____

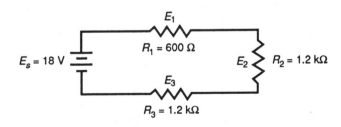

E_1

$R_1 = 600\ \Omega$

$E_s = 18\ \text{V}$ E_2 $R_2 = 1.2\ \text{k}\Omega$

E_3

$R_3 = 1.2\ \text{k}\Omega$

25. Find the total capacitance (C_t) in this circuit.   _____

   Note: C_t = branch 1 + branch 2

   branch 1 = $\dfrac{C_1(C_2 + C_3)}{C_1 + C_2 + C_3}$

   branch 2 = $\dfrac{C_4 C_5}{C_4 + C_5}$

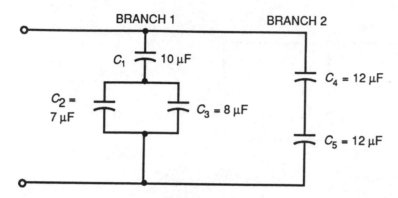

Unit 25   OHM'S LAW

BASIC PRINCIPLES OF OHM'S LAW

Ohm's law describes the relationship among voltage (*E*), current (*I*), and resistance (*R*). Georg Ohm found that if the voltage applied to a resistance was kept constant, and the resistance value was increased, the current through the resistance decreased. Ohm also found that for a constant resistance, when voltage was increased, current increased. These two findings can be expressed as follows.

1.   Current is *inversely* proportional to resistance. At a constant voltage, as the resistance increases the current decreases, and as the resistance decreases the current increases.

2.   Current is *directly* proportional to voltage. At a constant resistance, as the voltage increases the current increases, and as the voltage decreases the current decreases.

Expressing these relationships in a formula results in Ohm's law:

$E = IR$ where E = voltage in volts
I = current in amperes
R = resistance in ohms _____

To remember the other forms of Ohm's law, a wheel has been developed. Cover the value to be determined, and the other two values indicate the solution. To find current, cover the *I*; *E* over *R* (voltage divided by resistance) is the solution.

Example: A 2,700-ohm resistor is connected to 10 volts. What current will flow through the resistor?

$$I = \frac{E}{R} = \frac{10 \text{ V}}{2,700 \ \Omega} = 0.003\ 7 \text{ A} = 3.7 \text{ mA}$$

See Appendix Section II, Table III, for guide to multiplying and dividing units for Ohm's law.

PRACTICAL PROBLEMS

1. A 56-ohm resistor draws 12 milliamperes of current. Find the voltage drop across the resistor.   _____

2. When a 5-kilohm resistor is connected to a power supply, a current of 1.8 milliamperes is drawn. What is the output voltage?   _____

3. Find the current to the nearest tenth milliampere through the resistor in this circuit.   _____

$E_s = 6$ V   470 Ω

4. A 12-volt source supplies a current of 2.5 amperes. What is the resistance of the load?   _____

5. While checking a radio, a technician measures 5.2 volts across a 4.7-kilohm resistor. What current is passing through the resistor? (Express the answer to the nearer tenth milliampere.)   _____

6. Calculate the resistance of the diode in this circuit.   _____

$R_1 = 84.4 \, \Omega$

$I = 62.5$ mA

$E_s = 6$ V

D_1 $E_{D_1} = 0.7$ V

7. An unmarked battery has a 6.8-kilohm load attached and draws 1.6 milliamperes of current. Find the output voltage of the battery at this load.   _____

8. A rheostat is set to 600 ohms and connected to 24 volts. Find the current through the rheostat.   _____

9. Find the collector-to-emitter resistance (R_{ce}) in this transistor circuit.   _____

   Note: $R_{ce} = \dfrac{E_{ce}}{I_c}$

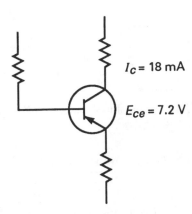

$I_c = 18$ mA

$E_{ce} = 7.2$ V

10. A radio has an antenna resistance of 1.2 ohms at a certain frequency. When a signal is received, a current of 3.8 microamperes is drawn. Find the signal voltage.   _____

11. The coil of a relay has a resistance of 180 ohms. Find the current when 6.3 volts are applied.   _____

12. What is the hot resistance of the bulb in this circuit?   _____

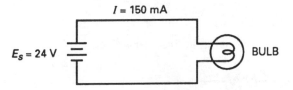

$I = 150$ mA

$E_s = 24$ V

BULB

13. What voltage is required to run an 8,200-ohm load that draws 500 microamperes of current?   _____

14. A transformer primary is connected to 120 volts. Find the current if the primary resistance is 480 ohms.   _____

15. The markings on a resistor have been removed. When it is connected to 10 volts, the current is 3.7 milliamperes. What is the resistor's value? Express the answer to the nearest ohm.   _____

16. A fuse blows when current exceeds its rated value. What minimum voltage causes the fuse in this circuit to blow?   _____

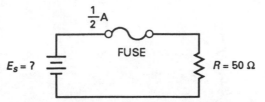

17. A capacitor is charged to 4.8 volts. What is the initial current when a 68-kilohm resistor is connected across the capacitor?   _____

18. A DC motor draws 1.2 amperes when connected to 48 volts. Find the motor's resistance to an external source.   _____

19. The photocell in this circuit produces 125 microamperes of current when connected to the load. Calculate the output voltage of the cell.   _____

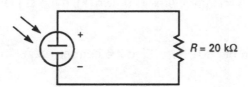

20. A 9-volt relay has a coil resistance of 200 ohms. How much current does it draw?   _____

21. Find the working resistance of the neon lamp in this circuit.   _____

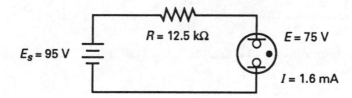

22. A diode has a resistance of 30 ohms when 21.6 milliamperes of current passes through it. What is the voltage drop across the diode?   _____

23. A rheostat is set to 4 kilohms. How much current does the battery in this
circuit supply to the rheostat?   _____

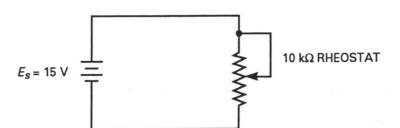

E_s = 15 V   10 kΩ RHEOSTAT

24. A load draws 1.5 amperes when connected to 6 volts. Find the load
resistance.   _____

25. A power source supplies 8 microamperes when a 2.5-megohm resistor is
connected in series with it. What is the voltage output?   _____

26. Complete this table.

	VOLTAGE (E)	CURRENT (I)	RESISTANCE (R)
a.	12 V		1.2 kΩ
b.		150 µA	4.2 kΩ
c.	2.7 V	1.8 mA	
d.	10 V		1.5 MΩ
e.		0.3 A	400 Ω
f.	280 mV	0.7 mA	
g.	2.2 kV		40 kΩ
h.		25 mA	0.1 kΩ
i.	5 V	20 µA	
j.	5 V		100 kΩ

 # Unit 26   POWER LAW

BASIC PRINCIPLES OF POWER LAW

The *Power law* states that voltage (*E*) times current (*I*) equals power (*P*).

$P = IE$ where   *P* = power in watts
                                 I = current in amperes
                                 E = voltage in volts

To remember the other forms of the Power law, use the wheel shown. Cover the value to be determined, and the other two values indicate the solution. To find voltage, cover the *E*; *P* over *I* (power divided by current) is the solution.

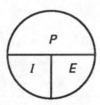

Example: A 500-ohm load is connected to a 25-volt power source. What is the consumption in watts? (First, Ohm's law must be applied to find current.)

$$I = \frac{E}{R} = \frac{25\ V}{500\Omega} = 0.05\ A$$

$$P = IE = 0.05\ A \times 25\ V = 1.25\ W$$

See Appendix Section II, Table III, for guide to multiplying and dividing units for Power law.

PRACTICAL PROBLEMS

1.   A load is connected to 12 volts and draws a current of 75 milliamperes. Find the power consumed.   _____

2. Find the current drawn by the load in this diagram.   _____

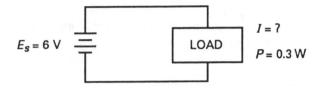

$E_s = 6$ V

LOAD

$I = ?$

$P = 0.3$ W

3. A 1½-watt load draws a current of 375 milliamperes. What is the applied
    voltage?   _____

4. A bulb conducts 0.25 ampere when connected to 120 volts. Find the power
    consumed.   _____

5. A wattmeter indicates a resistor is consuming 200 milliwatts of power. Use
    this circuit to determine the current.   _____

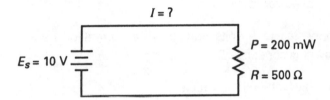

$I = ?$

$E_s = 10$ V

$P = 200$ mW

$R = 500 \ \Omega$

6. A 1-watt load is conducting 210 milliamperes of current. What is the value of
    the applied voltage? Express the answer to the nearest thousandth volt.   _____

7. The 2-watt resistor in this circuit is hot.

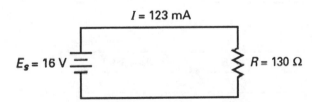

$I = 123$ mA

$E_s = 16$ V

$R = 130 \ \Omega$

    a. Is the resistor consuming more power than 2 watts?   a. _____

    b. Find the actual power consumed.   b. _____

8. A diode maintains a constant 0.7-volt drop across it when forward biased.
    Calculate the maximum current it can handle if it is a 1-watt diode. Round the
    answer to the nearest one-thousandth of an ampere.   _____

9. A 5-watt DC motor conducts 1.6 amperes. What is the applied voltage?   _____

10. The collector-to-emitter voltage of a power transistor is 7.4 volts. The transistor current is 85 milliamperes. Find the power consumed. _____

11. The 5-watt zener diode in the circuit maintains a constant voltage. Find the maximum current it can conduct without exceeding the wattage rating. _____

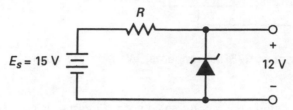

$E_s = 15$ V

R

+
12 V
–

12. A neon bulb is rated at 0.1 watt. When on, it conducts a current of 1.25 milliamperes. What is the applied voltage? _____

13. A radio antenna receives a signal of 4.8 millivolts that develops a current of 0.5 milliampere. Find the signal power at the antenna. _____

14. A 4.7-kilohm resistor is connected to 9 volts and draws a current of 1.9 milliamperes. How much power is consumed? _____

15. Find the power consumed in this high-voltage circuit. _____

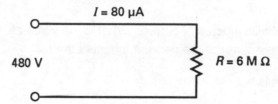

$I = 80$ μA

480 V

$R = 6$ M Ω

16. Complete this table.

	POWER (P)	VOLTAGE (E)	CURRENT (L)
a.	10 W	10 V	
b.		12 V	1.5 mA
c.	20 mW		5 mA
d.	3.6 W	1.8 V	
e.		0.5 V	0.03 mA
f.	7.3 W		0.2 A
g.	1.8 W	6 V	
h.		6 V	280 mA
i.	0.4 W		0.01 A
j.	31.5 W	9 V	

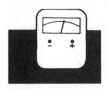

Unit 27 COMBINATION OHM'S LAW AND POWER LAW PROBLEMS

BASIC PRINCIPLES

The formulas for Ohm's law and the Power law can be rewritten to form many different equations. These equations are effectively formed into a wheel. When any two known values are used, either of the other two may be found.

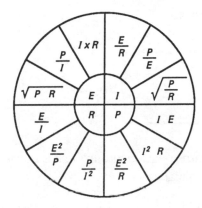

Examples:

a. A resistive load of 200 ohms is consuming 2 watts of power. What is the current flowing through the resistor?

$$I = \sqrt{\frac{P}{R}} = \sqrt{\frac{2\text{ W}}{200\Omega}} = 0.01 = 0.1\text{A}$$

b. A resistor is conducting a current of 1.2 milliamperes and is consuming 14.4 milliwatts of power. What is the value of the resistor?

$$R = \frac{P}{I^2} = \frac{14.4\text{ mW}}{(1.2\text{ mA})^2} = \frac{14.4 \times 10^{-3}}{(1.2 \times 10^{-3})^2} = \frac{14.4 \times 10^{-3}}{1.44 \times 10^{-6}} = 10{,}000\ \Omega \text{ or } 10\text{ k}\Omega$$

PRACTICAL PROBLEMS

1. A resistive load is consuming 2 watts of power. The applied voltage is 6 volts. Find the load resistance. _____

2. Find the power consumed by the load in this circuit.   _____

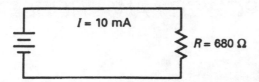

3. A 2-ohm resistor is consuming 8 watts of power. What is the current through
    the resistor?   _____

4. Calculate the source voltage in the circuit.   _____

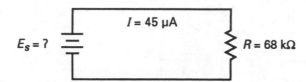

5. A resistor has a current of 20 milliamperes passing through it. It is consuming
    80 milliwatts of power. Compute the resistance.   _____

6. A 12-volt power supply has a 5-kilohm resistor connected to it. What is the
    power consumed by this resistor?   _____

7. What current is drawn by a 56-kilohm resistor with a 24-volt drop across it?
    Express the answer to the nearer microampere.   _____

8. A 40-ohm load consumes 2.5 watts of power. What voltage is applied?   _____

9. Find the resistance of the box in this diagram.   _____

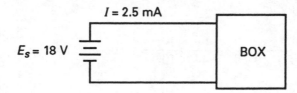

10. A transistor passes a current of 3.8 milliamperes with a voltage drop of 5 V.
    Find the power consumed.   _____

11. The zener diode in this schematic is consuming 0.81 watt. What is the current
     through the diode?   _____

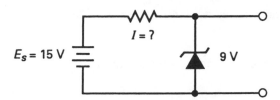

12. A solar cell produces 400 microamperes of current and 0.64 milliwatt of
     power to a load. Find the output voltage of the cell.   _____

13. A resistor is connected to a 36-volt power supply. It consumes 24 milliwatts
     of power. Find the resistor value.   _____

14. How much power is consumed by a 3.3-kilohm resistor with a 4.5-milliamp
     current through it? Express the answer to the nearest tenth milliwatt.   _____

15. Compute the current passing through the resistor in this circuit.   _____

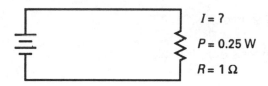

16. A load is supplied to a 17-milliampere current and consumes 0.34 watt of
     power. Find the applied voltage.   _____

17. A load is connected to a voltage source and draws 30 milliamperes of current.
     Find the load resistance if the power consumed is 90 milliwatts.   _____

18. Compute the power consumed by the resistor in this circuit. Express the
     answer to the nearest tenth milliwatt.   _____

19. A rheostat is set to 400 ohms. It is connected to 15 volts. Calculate the current.   _____

20. A 100-ohm resistor is consuming 1.44 watts. Find the value of the applied voltage.   _____

21. A resistive load draws 50 microamperes when connected to 12 volts.

    a. Find the load resistance.   a. _____

    b. Find the power consumption.   b. _____

22. Use this schematic.

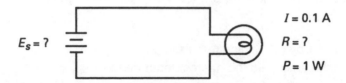

$I = 0.1$ A

$E_s = ?$

$R = ?$

$P = 1$ W

    a. Find the hot resistance of the bulb.   a. _____

    b. Find the value of the voltage source.   b. _____

23. The current through a 56-kilohm resistor is 150 microamperes.

    a. Calculate the voltage drop.   a. _____

    b. Calculate the power consumption.   b. _____

24. A 10-megohm resistor is placed across 120 volts.

    a. Find the current.   a. _____

    b. Find the power consumed.   b. _____

25. A load draws 0.25 ampere and consumes 0.5 watt.

    a. Find the load resistance.   a. _____

    b. Find the applied voltage.   b. _____

26. Complete this table.

	POWER (*P*)	VOLTAGE (*E*)	CURRENT (*I*)	RESISTANCE (*R*)
a.	1.5 W		50 mA	
b.		12 V		1.5 kΩ
c.	200 mW	5 V		
d.			400 µA	6.8 kΩ
e.	0.25 W			2,500 Ω
f.		0.7 V	17.5 mA	
g.	6.5 mW		0.2 mA	
h.		1.8 V		2 MΩ
i.			3.7 mA	10 kΩ
j.		10 V	10 mA	

Ratio and Proportion

Unit 28   RATIO

BASIC PRINCIPLES OF RATIO

Ratio

Ratio is the comparison of two numbers or quantities as a quotient. Ratios, like fractions, are generally reduced to lowest terms. The ratio 16:8 (read as "16 to 8") can be reduced to a ratio of 2:1 by dividing both numbers by 8. Ratios are also written in fractional form. The ratio 5:9 can also be written as

$$\frac{5}{9}$$

Inverse Ratio

An *inverse ratio* is the ratio in reverse order of the original ratio. The inverse ratio of 5:9 is 9:5 or

$$\frac{9}{5}$$

PRACTICAL PROBLEMS

Express these ratios in lowest terms.

1. 2:16 _____

2. 18:24 _____

3. 45:18 _____

4. $\frac{7}{8} : \frac{1}{16}$ _____

5. 6:9 _____

6. 7.2:1.6 _____

7. 1.6:7.2 _____

8. 120:8 _____

9. $\dfrac{1}{6}$: 6 _____

10. 1,024:256 _____

Find the ratio of the first quantity to the second quantity. Express the result in lowest terms.

Note: Quantities need to be expressed in identical units.

11. 2 V to 250 mV _____

12. 400 Ω to 1.6 kΩ _____

13. 120 mA to 20 µA _____

14. 0.5 V to 1.25 V _____

15. 220 Ω to 1.87 kΩ _____

16. Use this diagram to find the ratio of the meter resistance (R_m) to the shunt
    resistance (R_s). _____

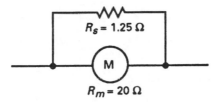

17. A quality control technician tests 822 radios and finds 12 defective radios.
    Find the ratio of acceptable radios to defective radios. _____

18. Find the ratio of primary turns to secondary turns in this transformer diagram. _____

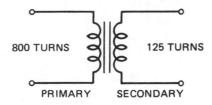

19. The input impedance of an amplifier is 60 kilohms. The output impedance is 1,200 ohms. Find the ratio of input impedance to output impedance.   _____

20. Find the ratio of R_1 to R_2 in this wheatstone bridge.   _____

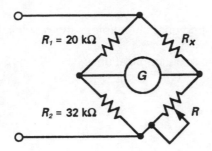

Unit 29   PROPORTION

BASIC PRINCIPLES OF PROPORTION

Proportion is the equality of two ratios.

$$5:2 = 25:10$$

There are two basic methods of solving problems dealing with proportion. One method is to multiply the *means* together and the *extremes* together. The means are the two inside terms, and the extremes are the two outside terms.

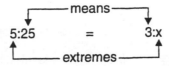

To solve this problem, multiply the means together and the extremes together.

$$5x = 75$$

To find the value of x, divide both sides by 5.

$$x = 15$$

Another method of solving this problem is to use *cross multiplication*. The two ratios are written as fractions separated by an equal sign when cross multiplication is used.

$$\frac{5}{25} = \frac{3}{x}$$

The top part of one ratio is then multiplied by the bottom part of the other.

$$\frac{5}{25} \times \frac{3}{x}$$

$$5x = 75$$

$$x = 15$$

Examples: Solve each proportion for the missing value:

a.
$$\frac{21}{3} = \frac{x}{9} \qquad \text{(original proportion)}$$
$$(21)\,(9) = (3)\,(x) \qquad \text{(cross-multiplication)}$$
$$189 = 3x$$
$$x = 63 \qquad \text{(solution)}$$

b.
$$\frac{y}{5.4} = \frac{9.1}{2.6} \qquad \text{(original proportion)}$$
$$(y)\,(2.6) = (5.4)\,(9.1) \qquad \text{(cross-multiplication)}$$
$$2.6y = 49.14$$
$$y = 18.9 \qquad \text{(solution)}$$

c.
$$\frac{10\ \text{k}\Omega}{4\ \text{k}\Omega} = \frac{26\text{k}\Omega}{z} \qquad \text{(original proportion)}$$
$$(10)\,(z) = (4)\,(26) \qquad \text{(cross-multiplication)}$$
$$10z = 104$$
$$z = 10.4\ \text{k}\Omega \qquad \text{(solution)}$$

PRACTICAL PROBLEMS

Find the missing quantity in each of these proportions.

1.   3:4 = 12:?   _____

2.   24 V:60 V = ?:500 V   _____

3.   0.675 A:? = 1.8 A:0.4 A   _____

4.   5.6 kΩ:0.2 kΩ = ?:10 kΩ   _____

5.   ?:3 = 18:4   _____

6.   56 Ω:14 Ω = ?:7 Ω   _____

7.   120¼:18½ = 143:?   _____

8. $\dfrac{28\ \text{mA}}{?} = \dfrac{4\ \text{mA}}{0.5\ \text{mA}}$ _____

9. $\dfrac{868}{24} = \dfrac{62}{?}$ _____

10. $\dfrac{?}{17 \text{ W}} = \dfrac{40 \text{ W}}{2.5 \text{ W}}$ _____

11. $\dfrac{120 \text{ V}}{0.25 \text{ V}} = \dfrac{48 \text{ V}}{?}$ _____

12. $\dfrac{15}{?} = \dfrac{110}{385}$ _____

13. $\dfrac{0.01 \text{A}}{0.002 \text{ A}} = \dfrac{?}{0.5 \text{ A}}$ _____

14. $\dfrac{10 \text{ k}\Omega}{4 \text{ k}\Omega} = \dfrac{?}{6 \text{ k}\Omega}$ _____

15. Three cable television installers complete a section of 10 houses in 6 hours. How many hours will it take five installers, working at the same rate, to complete this same section of houses? (Hint: First find the number of houses three workers complete in 1 hour.) _____

Note: Use this transformer diagram and these formulas for problems 16–18.

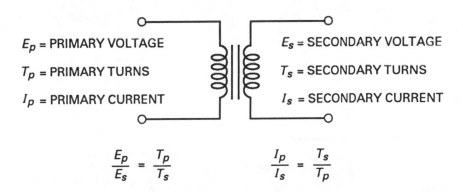

E_p = PRIMARY VOLTAGE E_s = SECONDARY VOLTAGE

T_p = PRIMARY TURNS T_s = SECONDARY TURNS

I_p = PRIMARY CURRENT I_s = SECONDARY CURRENT

$$\frac{E_p}{E_s} = \frac{T_p}{T_s} \qquad\qquad \frac{I_p}{I_s} = \frac{T_s}{T_p}$$

16. A transformer has a 400-turn primary and a 50-turn secondary. Find the secondary voltage if 120 volts are applied. _____

17. A transformer supplies a secondary current of 0.8 ampere. The primary turns are 150 and the secondary turns are 600. Find the primary current. _____

18. A 1,200-turn primary has an applied voltage of 60 volts. Find the number of turns in the secondary if its voltage is 24 volts. _____

19. The shunt resistance (R_s) in this circuit is designed to pass 0.9 milliampere. Use this circuit and formula to determine the value of R_s. Express the answer to the nearest hundredth.

    _____

    Note: $\dfrac{R_s}{R_m} = \dfrac{I_m}{I_s}$

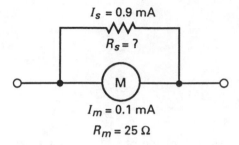

$I_s = 0.9$ mA

$R_s = ?$

M

$I_m = 0.1$ mA

$R_m = 25\ \Omega$

20. A roll of number 20 AWG copper wire is 3,000 feet long and has a total resistance of 30.45 Ω. Find the amount of resistance in 750 feet. Express the answer to the nearest hundredth.

    _____

21. A piece of wire 1,000 feet long has a resistance of 6.4 ohms. How long is a piece of wire with a resistance of 22.4 ohms?

    _____

Note: Use this wheatstone bridge and balanced circuit formula in problems 22–24.

$$\frac{R_1}{R_2} = \frac{R_x}{R}$$

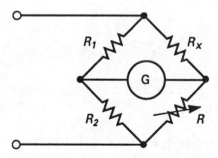

R_1 R_x

G

R_2 R

22. $R_1 = 10$ kΩ, $R_2 = 40$ kΩ, and $R = 30$ kΩ. Find the value of R_x.

    _____

23. R_1 and $R_2 = 25$ kΩ, and $R = 5.6$ kΩ. Find the value of R_x.

    _____

24.   The ratio of R_1 to R_2 is 5. Find the value of R_x if $R = 20$ kilohms.   _____

25.   The antenna shown is a Yaggie type. The dimensions given are for an antenna designed to have maximum gain at a frequency of 50.75 megahertz. Find the dimensions to make a Yaggie-type antenna that will have a maximum gain at 605 megahertz.

   Note: Frequency and length are inversely proportional.

   a.   _____

   b.   _____

   c.   _____

   d.   _____

   e.   _____

   f.   _____

   g.   _____

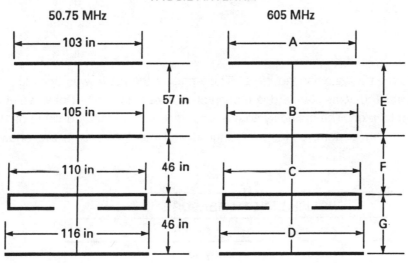

YAGGIE ANTENNA

Measurement and Graphs

UNIT 30 LINEAR MEASURE

BASIC PRINCIPLES OF LENGTH MEASURE

The English System of Measure

The English system of measure was originally based on the measurements of the king's body and other natural things. The length of the king's foot, for example, was the standard length of a foot. The middle joint of the king's index finger was the standard length of one inch. Another weight measurement still used today is the grain. A grain was the weight of a grain of wheat.

These measurements are impossible to duplicate because no two things that occur in nature are exactly the same. No two grains of wheat weigh exactly the same, for instance. Today, there are standards of measurement that are kept by the International Bureau of Weights and Standards. The length of a foot, for example, is always the same.

The Metric System

The standard length in the metric system is the *meter.* The length of the meter was originally based on a measurement of the earth. Today, its standard is the length of a certain number of wavelengths in a ray emitted by the element krypton. The metric system is also based on a standard value of 10. This is an advantage over the English system. Lengths of English and metric measure are shown in the following charts.

ENGLISH LENGTH MEASURE		
12 inches (in)	=	1 foot (ft)
3 feet (ft)	=	1 yard (yd)
1,760 yards (yd)	=	1 mile (mi)
5,280 feet (ft)	=	1 mile (mi)

METRIC LENGTH MEASURE		
10 millimeters (mm)	=	1 centimeter (cm)
10 centimeters (cm)	=	1 decimeter (dm)
10 decimeters (dm)	=	1 meter (m)
10 meters (m)	=	1 dekameter (dam)
10 dekameters (dam)	=	1 kilometer (km)

ENGLISH-METRIC EQUIVALENTS LENGTH MEASURE	
1 inch (in)	$\approx$ 25.4 millimeters (mm)
1 inch (in)	$\approx$ 2.54 centimeters (cm)
1 foot (ft)	$\approx$ 0.3048 meter (m)
1 yard (yd)	$\approx$ 0.9144 meter (m)
1 mile (mi)	$\approx$ 1.609 kilometers (km)
1 millimeter (mm)	$\approx$ 0.03937 inch (in)
1 centimeter (cm)	$\approx$ 0.39370 inch (in)
1 meter (m)	$\approx$ 3.28084 feet (ft)
1 meter (m)	$\approx$ 1.09361 yards (yd)
1 kilometer (km)	$\approx$ 0.62137 mile (mi)

Note: $\approx$ indicates "approximately equals"

Many measurements in electronics involve linear measure (inches, feet, centimeters, meters, and so on), and the technician may need to convert from one unit of linear measure to another. Units of measure are similar to numbers in that feet $\times$ feet = feet^2 or square feet. When converting from one unit of linear measure to another, caution must be taken that the units of measure are converted correctly. The primary rule for performing conversions among units of linear measure is:

Always begin the conversion by setting up the units. Only then should numbers be inserted. Never give a measurement without indicating the unit of measurement.

Example: Convert 18 miles (mi) to kilometers (km).

1.   The first step is to set up the conversion to produce the correct units of measurement, starting with the unit of the given measurement:

$$\text{mi} \times \frac{\text{km}}{\text{mi}} = \text{km}$$

Note that in this step, the miles on the left side of the equation cancel, leaving kilometers, which means step two can now be done.

2.   Insert the numbers into the conversion. In looking at the table, there are two conversions involving miles and kilometers. The first indicates that 1 mile = 1.609 kilometers. The numbers can now be put into the conversion formula:

$$18 \text{ mi} \times \frac{1.609 \text{ km}}{1 \text{ mi}} = (18 \times 1.609) \text{ km} = 28.962 \text{ km}$$

The other conversion involving kilometers and miles indicates that 1 km = 0.621 4 mile. Inserting these numbers in the conversion formula should produce the same results.

$$18 \text{ mi} \times \frac{1 \text{ km}}{0.621 \text{ 4 mi}} = \left(\frac{18}{0.621 \text{ 4}}\right) \text{km} = 28.967 \text{ km}$$

Note that in this example, the two answers are a bit different. This is due to the rounding of the conversion values in the table. However, they are the same at two decimal places.

Example: Convert 145.82 centimeters (cm) to feet (ft). Round the answer to two decimal places.

$$145.82 \text{ cm} \times \frac{1 \text{ in}}{2.4 \text{ cm}} \times \frac{1 \text{ ft}}{12 \text{ in}} = \frac{145.82}{(2.54)(12)} \text{ft} = 4.78 \text{ ft}$$

PRACTICAL PROBLEMS

Express each linear measurement as indicated. Answers expressed as decimals should be rounded to two decimal places.

1.   2 miles as kilometers   _____

2.   12 feet as inches   _____

3.   6.2 meters as feet   _____

4.   3.5 miles as inches   _____

5.   152 centimeters as millimeters   _____

6.   16½ inches as centimeters   _____

7.   1.2 kilometers as feet   _____

8.   820 centimeters as meters   _____

9.   6 feet 8½ inches as inches   _____

10.   2 yards 1 foot 3¾ inches as meters   _____

11.   A section of coaxial cable is 27 feet long. What is this length in inches?   _____

12.   The length of a power resistor is 2½ inches. Express this length in
      centimeters.   _____

13.   An antenna is 24 feet 6¾ inches high. Find the total height in inches.   _____

14.   A delivery van travels 80 kilometers. What is the distance in miles?   _____

15.   Fifteen television consoles are lined up end to end. Each television is 3 feet
      2½ inches long. Find the total length in feet and inches.   _____

16.   Find the height in millimeters of a transistor that is ¼ inch high.   _____

17.   A radio station transmits a distance of 40 miles. Express this distance in
      kilometers.   _____

18.   An indoor FM antenna is 7½ feet in length. Express this length in yards.   _____

19.   A part is 4¾ inches tall. What is its height in millimeters?   _____

20.   A television picture tube measures 21 inches diagonally. Find this length in
      millimeters.   _____

21.   The diameter of a turntable platter is 35.56 centimeters. Express this
      diameter in inches.   _____

22.   An extension cord is 9 feet long. Find this length in centimeters.   _____

23. Light travels approximately 55,800 miles in 0.3 second. Express in kilometers the distance light travels in one second.   _____

24. A tape recorder moves the audio tape at a speed of 7½ inches per second. What is the speed in centimeters per second?   _____

25. How many feet of wire are wound on a 100-meter roll?   _____

Unit 31 SURFACE AND VOLUME MEASURE

BASIC PRINCIPLES OF SURFACE AND VOLUME MEASURE

Surface measure is often referred to as *area* measure. Area measurements are two-dimensional (length × width) and describe the surface area of an object. They are expressed as squared quantities, such as square inches (in^2), square feet (ft^2), square meters (m^2), and so on.

Example: Find the area of an object with a length of 3 feet and a width of 2 feet.

$$3 \text{ ft} \times 2 \text{ ft} = 6 \text{ ft}^2$$

Volume measurements are three-dimensional (length × width × height) and are expressed as cubic quantities, such as cubic inches (in^3), cubic feet (ft^3), cubic meters (m^3), and so on.

Example: How many gallons of water can be contained in a tank that measures 4 feet by 6 feet by 2 feet?

Note: One cubic foot will hold 7.5 gallons of water.

$$4 \text{ ft} \times 6 \text{ ft} \times 2 \text{ ft} - 48 \text{ ft}^3$$
$$48 \text{ ft}^3 \times \frac{7.5 \text{ gal}}{\text{ft}^3} = 360 \text{ gal}$$

PRACTICAL PROBLEMS

1. Find the area (*A*) of the plate in the diagram. _____

   Note: $A = lw$

   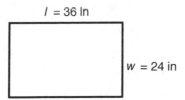

   $l = 36$ In

   $w = 24$ in

2. The chip within an integrated circuit is square in shape with one edge 4 millimeters long. Find the area (*A*) of the chip. _____

   Note: $A = s^2$

3. A television station broadcasts a signal to an area shaped like an ellipse. Use this diagram to compute the broadcast area (*A*).

   Note: $A = \pi ab$, where $\pi = 3.1416$

   _____

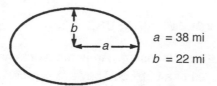

   $a = 38$ mi

   $b = 22$ mi

4. A computer service center is quoted a price of $38 per square foot for building an addition. If the addition is 42 feet by 60 feet, find the cost.

   _____

5. A speaker is shaped like a triangle. Use this diagram to determine the front surface area of the speaker.

   Note: $A = \frac{1}{2}\, bh$

   _____

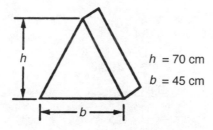

   $h = 70$ cm

   $b = 45$ cm

6. The top surface of a television console measures 3 feet 4 inches by 18 inches. Find the surface area in square inches.

   _____

7. A radio is shaped like a cube. Find the volume (*V*) in cubic inches if one edge is 4 inches long.

   Note: $V = s^3$

   _____

8. The base of a transmitting tower measures 24 feet 6 inches by 28 feet 6 inches. Compute the area of the base in square feet. (Hint: Express the inches as feet.)

   _____

9. Find the entire surface area (*A*) of this stereo receiver.   _____

    Note: *A* = 2*ab* + 2*bc* + 2*ac*

    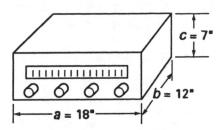

10. The base of a transformer measures 7.2 centimeters by 5.6 centimeters. How much area on a chassis plate is covered when the transformer is mounted?   _____

11. A potentiometer is circular and has a diameter (*d*) of 1 inch. Compute the circumference (*C*) of the potentiometer.   _____

    Note: $C = \pi d$, where $\pi = 3.1416$

12. A printed circuit board measures 6 inches by 3½ inches. Another circuit board is 5¾ inches long and 3 inches wide. Find the difference in surface area between the two boards.   _____

13. Calculate the volume (*V*) to the nearer cubic inch of the cylindrical tank in this diagram.   _____

    Note: $V = \pi r^2 h$, where $\pi = 3.1416$

    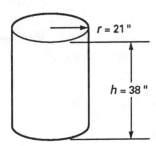

14. A speaker is circular and has a radius of 4 inches. Find the surface area (*A*).   _____

    Note: $A = \pi r^2$, where $\pi = 3.1416$

15. A radio station has a weather balloon with a radius (*r*) of 30 centimeters. Calculate the volume (*V*) of the balloon.   _____

    Note: $V = \frac{4}{3}\pi r^3$, where $\pi = 3.1416$

16. A circuit board has an area of 29¾ square inches. The width is 4¼ inches. Find the length.

_____

17. A rectangular etching tank has a base that measures 4 feet by 2 feet. How many cubic feet are added if the height is changed from 3 feet to 3.5 feet?

_____

18. The container in this diagram is ¼ full. What is the volume, in cubic inches, of the filled portion?

_____

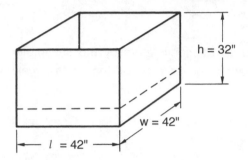

h = 32"

w = 42"

l = 42"

19. A square chassis plate has an area of 64 square inches. Find the length of one side.

_____

20. One type of circuit board material costs $4.00 per square foot. Find the cost of a piece 42 inches by 18 inches.

_____

21. A solar collector 18 feet long and 8 feet high is used to heat a tank of water. The tank is 3 feet by 3 feet by 4 feet. The amount of heat received from the sun is approximately 100 watts per square foot. A BTU (British Thermal Unit) is the amount of heat needed to raise the temperature of one pound of water one degree Fahrenheit. One cubic foot will hold 7.5 gallons of water, and water weighs 8.34 pounds per gallon. One watt will produce 3.41 BTUs of heat energy per hour.

a. How much heat in watts does the solar collector take in?

a. _____

b. How much does the water in the tank weigh?

b. _____

c. How many BTUs are needed to raise the temperature of the water 10°F?

c. _____

d. Assuming the solar collector is 65% efficient, how many BTUs does the collector produce in one hour?

d. _____

e. How long will it take the collector to raise the water temperature 10°F?

e. _____

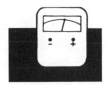

Unit 32　GRAPHS

BASIC PRINCIPLES OF GRAPHS

A *graph* is a picture showing the relationship between two or more quantities. There are many types of graphs, including the bar graph, circle graph, and coordinate graph. The graphs presented in this unit are coordinate graphs. Two axes and four quadrants make up the coordinate graph.

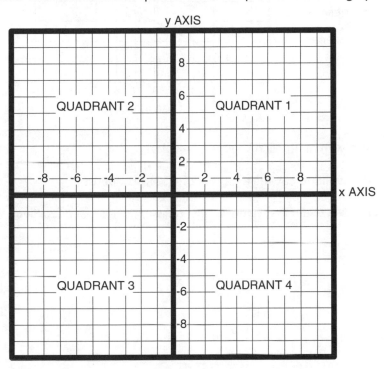

The graph is divided into four sections, or quadrants, by a horizontal line (*x* axis) and a vertical line (*y* axis). Note that the origin is the intersection of the *x* and *y* axes and that positive and negative numbers extend out from the origin on both axes. This unit interprets elements in the first and fourth quadrants of graphs.

To locate the *y* value of a point on a graph, locate the *x* value given in a problem on the *x* axis and extend a vertical line until it intersects the graph. Extend a horizontal line from this intersection point to the left until it intersects the *y* axis.

To locate both *x* and *y* values on the axes, reverse the procedure by extending a horizontal line from a given point on the graph to read the *y* axis and a vertical line to read the *x* axis.

PRACTICAL PROBLEMS

This graph demonstrates how a fixed resistance reacts as different voltages are applied. Notice that as the applied voltage increases, the current also increases. Use this graph for problems 1–4. In this graph, the *x* axis holds values of voltage (*E*) and the *y* axis holds values of current (*I*).

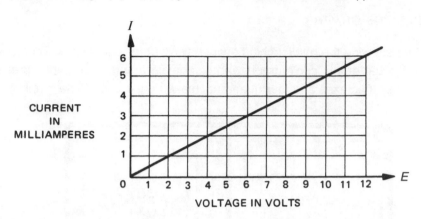

1. What current exists with an applied voltage of 4 volts?   _____

2. Approximately what voltage causes a current of 4.5 milliamperes?   _____

3. Using Ohm's law, calculate the value of the resistance for each applied voltage.

    Note: $R = \dfrac{E}{I}$

    a. 4 V   a. _____

    b. 8 V   b. _____

    c. 12 V   c. _____

4. Using the Power law, calculate the power consumed for each applied voltage.

    Note: $P = IE$

    a. 6 V   a. _____

    b. 10 V   b. _____

A diode conducts a small amount of current until it is forward biased past a certain point. When Ohm's law is applied to this graph, it is seen that at a certain voltage, the resistance of the diode drops and current increases rapidly. Use this graph for problems 5–9.

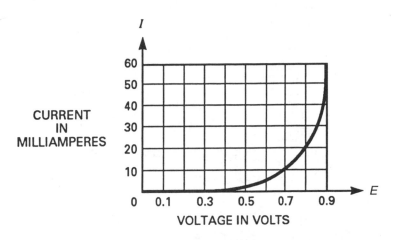

CURRENT
IN
MILLIAMPERES

VOLTAGE IN VOLTS

5. Approximately what current exists with a 0.1 voltage drop?   _____

6. Use Ohm's law to calculate the resistance of the diode at 0.7 volt.   _____

    Note: $R = \dfrac{E}{I}$

7. Use Ohm's law to calculate the resistance of the diode at 0.9 volt.   _____

8. Use the Power law to calculate the power consumed at 0.7 volt.   _____

    Note: $P = IE$

9. Use the Power law to calculate the power consumed at 0.8 volt.   _____

This graph shows a family of characteristic curves for a collector-emitter transistor circuit. The graph indicates that for a constant value of base current (I_b), the collector current (I_c) (represented on the y axis) remains relatively constant. Notice that changes in collector-emitter voltage (E_{ce}) (represented on the x axis) have little effect upon collector current. This family of curves is based upon six different values of base current. Use this graph for problems 10–15.

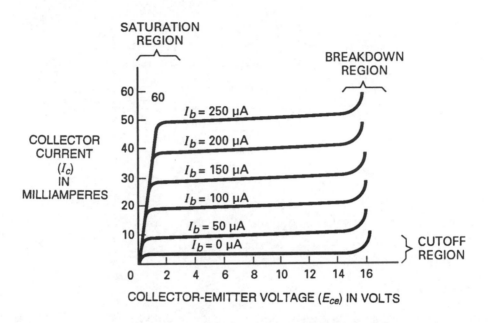

10. What is the value of I_c for a base current of 100 microamperes (µA)? _____

11. A change of 50 microamperes of base current causes how much of a change in I_c? _____

12. Calculate the approximate resistance from collector to emitter for E_{ce} = 10 volts and I_b = 250 microamperes. _____

    Note: $R_{ce} = \dfrac{E_{ce}}{I_c}$

13. Calculate the approximate resistance from collector to emitter for E_{ce} = 10 volts and I_b = 50 microamperes. _____

14. Compute the current gain (β) of the circuit for I_b = 100 microamperes. _____

    Note: $\beta = \dfrac{I_c}{I_b}$

15. Calculate the current gain (β) of the circuit for I_b = 250 microamperes. _____

This graph displays the time required for a capacitor to charge to 40 volts. Use this graph to answer problems 16–18.

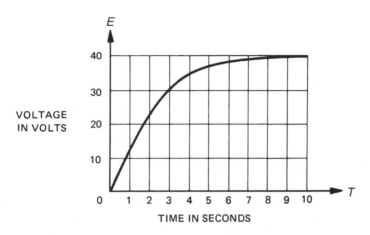

16.   How many seconds does it take for the capacitor to charge to 40 volts?   _____

17.   Find the approximate voltage the capacitor is charged to after 3 seconds.   _____

18.   How many seconds does it take the capacitor to receive a charge of 20 volts?   _____

A filter will pass some voltages and block others. Use this graph for problems 19–21.

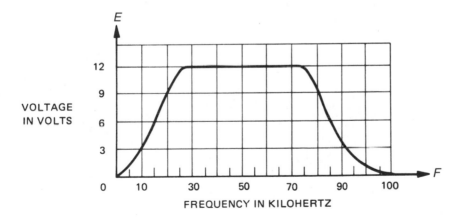

19.   At what frequency (or frequencies) is the output approximately 6 volts?   _____

20.   What is the output voltage at 50 kilohertz?   _____

21.   At what frequency does the output start to decrease?   _____

A sine wave of alternating voltage increases and decreases over 360° of rotation. Use this graph for problems 22–25.

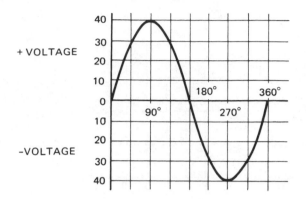

22. Find the voltage at 270 degrees.   _____

23. At approximately which degree points is the output 20 volts?   _____

24. Find the output voltage at 180 degrees.   _____

25. What is the approximate output voltage at 225 degrees?   _____

Percentages, Averages, and Tolerances

SECTION

8

Unit 33   *PERCENTAGES AND AVERAGES*

BASIC PRINCIPLES OF PERCENTAGES AND AVERAGES

Percent begins with a special kind of fraction, one that has 100 in the denominator. Percent means the number of parts per 100, and so a fraction such as $^{63}/_{100}$ represents 63 parts out of a maximum of 100, or 63%.

When the division indicated by the fraction is completed, the result is a decimal fraction. The result of dividing 63 by 100, for example, is 0.63 (or 63 hundredths). *A decimal can be converted to a percent by moving the decimal point two places to the right.* Thus, 0.63 is the same as 63 percent or 63%. The number 0.09 is equal to 9%, and 0.375 is 37.5%. The following rules apply to working with percentages.

To change a fraction to a percent, perform the division indicated by the fraction. This will result in a decimal. Move the decimal point two places to the right to obtain the percent. Put % after the number.

Example: Convert ¾ to a percent.

$$\frac{3}{4} = 3 \div 4 = 0.75 = 75\%$$

To change a percent to a decimal, remove the percent symbol and move the decimal point two places to the left.

Example: Convert 87.4% to a decimal.

$$87.4\% = 0.874$$

> To change a percent to a fraction, remove the percent symbol and divide by 100. Reduce the fraction to lowest terms.

Example: Convert 65% to a fraction.

$$65\% = \frac{65}{100} = \frac{13}{20}$$

> To find the percentage of a given number, convert the percent to its decimal equivalent and then multiply the decimal equivalent by the given number. The word *of* indicates multiplication in this case.

Example: What is 37% of 85?

$$37\% = 0.37 \text{ so that } 0.37 \times 85 = 31.45$$

> To determine the relationship one number has with another in terms of a percent, divide. The number following the word *of* is the divisor. Then move the decimal point two places to the right and put % after the number.

Example: 17 is what percent of 85?

$$\frac{17}{85} = 0.2 = 20\%$$

The *average* of a series of numbers is the one number that best represents all the numbers in the series. For example, the average hourly wage for a technician may be computed. On one day, the technician may earn more than the average, and on another day less. Follow this rule to compute average.

> Add all the numbers in the series. Divide the total by the number of numbers in the series. The result is the average.

Example: What is the average of 8 volts, 10 volts, 9 volts, 12 volts, and 7.5 volts?

$$8 \text{ V} + 10 \text{ V} + 9 \text{ V} + 12 \text{ V} + 7.5 \text{ V} = 46.5 \text{ V}$$

$$\frac{46.5 \text{V}}{5} = 9.3 \text{ V}$$

PRACTICAL PROBLEMS

Find these percentages.

1. 10% of 75 _____

2. 55% of 138 _____

3. 18½% of 220 _____

Find these averages. Round the answers to three places if necessary.

4. 10, 18, 7, 9, 6 _____

5. 0.27, 1.05, 0.95, 0.88 _____

6. 13.3, 27.2, 18.9, 16.5 _____

7. A test technician checks 22 circuit boards out of a total of 40. What percent of them is tested? _____

8. A power supply can be varied from 0 to 60 volts. What percent of the total output is 21 volts? _____

9. A battery is designed to provide 24 volts. During six tests, the output voltages are measured at 23.8 volts, 24.12 volts, 24.1 volts, 23.92 volts, 23.56 volts, and 24.08 volts. Find the average output. _____

10. The average output (E_{avg}) of a power supply is 12.5 volts. The rms ripple voltage is 0.05 volt. Find the percent of ripple. _____

    Note: % of ripple $= \dfrac{E_{rms}}{E_{avg}} \times 100\%$

11. During a four-week period, a technician earns $627, $660, $627, and $660. If the rate of pay is $16.50 per hour, find the average hours worked per week. _____

12. A developing solution for circuit boards contains 1.2 liters of water. To this is added 0.8 liter of solution A and 0.5 liter of solution B. What percent of the total liters is solution A? _____

13.   With no load attached, the no-load voltage (E_{nl}) of a power supply is 12 volts. When a 100-ohm load is attached, the full-load voltage (E_{fl}) is 11.7 volts. Calculate the percent of *regulation* to the nearer hundredth.   _____

   Note: % of regulation $= \dfrac{E_{nl} - E_{fl}}{E_{fl}} \times 100\%$

14.   An electronics kit is designed to be put together in 44 hours. The actual number of hours it takes to assemble it is 15% less than expected. How long does the assembly take?   _____

15.   Jackie says she averages 7.5 minutes per board while assembling components on circuit boards. How many boards does she complete in 45 minutes?   _____

16.   A roll of wire is 500 feet long. What percent remains after 125 feet are cut off?   _____

17.   A resistor is operating at 80% of its maximum power rating. If it is consuming 4 watts of power, find its maximum rating.   _____

18.   The regular price for a video game is $56. What is the sale price during a 25% off sale?   _____

19.   During an experiment five photovoltaic cells are tested. The output voltages are 1.82 volts, 2.08 volts, 2.00 volts, 1.91 volts, and 2.14 volts. Find the average output.   _____

20.   The weekly salary for a technician is $680.00. Find the percent of salary remaining if deductions total $245.90.   _____

21.   The input power to a transformer is 46 watts. The power delivered to the secondary is 42 watts. What percent to the nearer tenth of the power is lost?   _____

Unit 34   *TOLERANCES AND COLOR CODE*

BASIC PRINCIPLES OF TOLERANCES AND COLOR CODE

Composition carbon, carbon film, and metal film resistors are generally marked with bands of color that indicate resistance value and tolerance. Each color represents a numerical value or percent tolerance. The colors and corresponding numerical values are as follows:

Black	0
Brown	1
Red	2
Orange	3
Yellow	4
Green	5
Blue	6
Violet	7
Gray	8
White	9

Resistors that have a tolerance of plus or minus (+/−) 20% will be marked with three bands of color. Resistors that have a tolerance of plus or minus 10%, 5%, or 2% will be marked with four bands of color, and resistors with a tolerance of plus or minus 1% will be marked with five bands of color.

Tolerance

Tolerance indicates the range of resistance from its marked value that a resistor can have and still be within its rating.

Example: Assume that a resistor has a marked value of 33,000 ohms and a tolerance of 10%. To determine if this resistor is within its rating, find 10% of 33,000.

$$33,000 \times 0.10 = 3000$$

To determine if the resistor is within its rating, add and subtract 3,000 from 33,000. The resistor will be within its rating if its measure value falls between 36,000 (33,000 + 3,000) and 30,000 (33,000 − 3,000) ohms.

Reading Resistor Values

To understand how to read the value of resistance and tolerance, refer to the resistor color code chart.

Resistor Color Code Chart

Color	1st Band	2nd Band	3rd Band	4th Band
	1st Digit	2nd Digit	Multiplier	Tolerance
Black	–	0	0	–
Brown	1	1	1	1%
Red	2	2	2	2%
Orange	3	3	3	–
Yellow	4	4	4	–
Green	5	5	5	–
Blue	6	6	6	–
Violet	7	7	7	–
Gray	8	8	8	–
White	9	9	9	–
Gold	–	–	x0.1	5%
Silver	–	–	x0.01	10%
None	–	–	–	20%

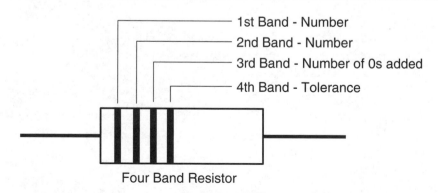

1st Band - Number
2nd Band - Number
3rd Band - Number of 0s added
4th Band - Tolerance

Four Band Resistor

Most resistors contain four bands of color. The first two bands represent numbers or digits. The third band is called the multiplier band. This band indicates the number of zeros that should be added to the first two numbers. The fourth band is the tolerance band. Resistors that have only three bands or color have a tolerance of +/– 20%.

Example: Determine the value and tolerance of the resistor shown in the following figure.

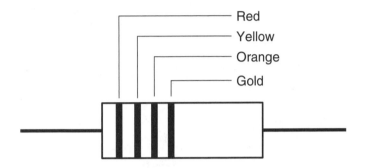

Solution: The first band is red, which indicates the number 2. The second band is yellow, which indicates 4. The third band is orange, which indicates 3. Since this is the multiplier band, add three zeros to the first two numbers. The value of the resistors is 24,000 ohms. The fourth band is gold, which indicates a tolerance of +/– 5%.

Standard Resistance Values

Not all values of resistance are available in composition resistors. The standard resistance values chart lists the standard base values for different tolerances. Notice that in the 2%, 5% column there are more standard values listed than in the 10% column. All of these base values can be multiplied or divided by factors to 10 to produce other values. In the 2%, 5% column, one of the standard values is 22. Resistors with values of 0.22 ohm, 02.2 ohms, 22 ohms, 220 ohms, 2,200 ohms, 22,000 ohms, 220,000 ohms, and 2,200,000 ohms can all be found in the tolerance range of 2% or 5%. Do not try to find a 25,000-ohm resistor in this tolerance range, however, because it doesn't exist.

STANDARD RESISTANCE VALUES

.1%, .25% .5%	1%	.1%, .25% .5%	1%	.1%, .25% .5%	1%	.1%, .25% .5%	1%	.1%, .25% .5%	1%
10.0	10.0	17.2	-	29.4	29.4	50.5	-	86.6	86.6
10.1	-	17.4	17.4	29.8	-	51.1	51.1	87.6	-
10.2	10.2	17.6	-	30.1	30.1	51.7	-	88.7	88.7
10.4	-	17.8	17.8	30.5	-	52.3	52.3	89.8	-
10.5	10.5	18.0	-	30.9	30.9	53.0	-	90.9	90.9
10.6	-	18.2	18.2	31.2	-	53.6	53.6	92.0	-
10.7	10.7	18.4	-	31.6	31.6	54.2	-	93.1	93.1
10.9	-	18.7	18.7	32.0	-	54.9	54.9	94.2	-
11.0	11.0	18.9	-	32.4	32.4	55.6	-	95.3	95.3
11.1	-	19.1	19.1	32.8	-	56.2	56.2	96.5	-
11.3	11.3	19.3	-	33.2	33.2	56.9	-	97.6	97.6
11.4	-	19.6	19.6	33.6	-	57.6	57.6	98.8	-
11.5	11.5	19.8	-	34.0	34.0	58.3	-		
11.7	-	20.0	20.0	34.4	-	59.0	59.0		
11.8	11.8	20.3	-	34.8	34.8	59.7	-		
12.0	-	20.5	20.5	35.2	-	60.4	60.4		
12.1	12.1	20.8	-	35.7	35.7	61.2	-		
12.3	-	21.0	21.0	36.1	-	61.9	61.9		
12.4	12.4	21.3	-	36.5	36.5	62.6	-		
12.6	-	21.5	21.5	37.0	-	63.4	63.4		
12.7	12.7	21.8	-	37.4	37.4	64.2	-		
12.9	-	22.1	22.1	37.9	-	64.9	64.9		
13.0	13.0	22.3	-	38.3	38.3	65.7	-		
13.2	-	22.6	22.6	38.8	-	66.5	66.5		
13.3	13.3	22.9	-	39.2	39.2	67.3	-		
13.5	-	23.2	23.2	39.7	-	68.1	68.1		
13.7	13.7	23.4	-	40.2	40.2	69.0	-		
13.8	-	23.7	23.7	40.7	-	69.8	69.8		
14.0	14.0	24.0	-	41.2	41.2	70.6	-		
14.2	-	24.3	24.3	41.7	-	71.5	71.5		
14.3	14.3	24.6	-	42.2	42.2	72.3	-		
14.5	-	24.9	24.9	42.7	-	73.2	73.2		
14.7	14.7	25.2	-	43.2	43.2	74.1	-		
14.9	-	25.5	25.5	43.7	-	75.0	75.0		
15.0	15.0	25.8	-	44.2	44.2	75.9	-		
15.2	-	26.1	26.1	44.8	-	76.8	76.8		
15.4	15.4	26.4	-	45.3	45.3	77.7	-		
15.6	-	26.7	26.7	45.9	-	78.7	78.7		
15.8	15.8	27.1	-	46.4	46.4	79.6	-		
16.0	-	27.4	27.4	47.0	-	80.6	80.6		
16.2	16.2	27.7	-	47.5	47.5	81.6	-		
16.4	-	28.0	28.0	48.1	-	82.5	82.5		
16.5	16.5	28.4	-	48.7	48.7	83.5	-		
16.7	-	28.7	28.7	49.3	-	84.5	84.5		
16.9	16.9	29.1	-	49.9	49.9	85.6	-		

2%,5%	10%
10	10
11	-
12	12
13	-
15	15
16	-
18	18
20	-
22	22
24	-
27	27
30	-
33	33
36	-
39	39
43	-
47	47
51	-
56	56
62	-
68	68
75	-
82	82
91	-

Notice the number of standard values listed in the 1% column. Also notice that the standard values listed contain three numbers instead of two. This is the reason that resistors with a tolerance of +/– 1% must contain five bands of color instead of four.

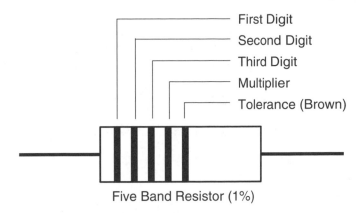

Five Band Resistor (1%)

The first three bands indicate number values. The fourth band is the multiplier, and the fifth band, which will always be brown on a 1% resistor, indicates tolerance.

Example: The resistor shown here contains five bands of color. What is the resistance value and tolerance?

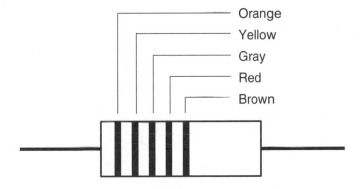

Solution: The first band is orange, which is 3. The second band is yellow, which is 4. The third band is gray, which is 8. The fourth band is red, which is 2. Since the fourth band is the multiplier band, add two zeros to the first three digits. The resistance value is 34,800 ohms. The fifth band of brown indicates a tolerance of +/– 1%.

Examples: Solve the following color code problems.

a.   Find the value of a resistor with the colors red, yellow, orange, and gold. According to the chart, the first digit is 2, the second digit is 4, followed by 3 zeros. This would indicate a value of 24,000 ohms. The fourth band is gold and indicates a 5% tolerance. This means the actual value may fall between 22,800 ohms and 25,200 ohms.

b.   Find the color code of a 1-kilohm resistor with a 10% tolerance. Since 1 kilohm = 1000 ohms, reference to the chart shows that the colors would be brown, black, red, and silver.

c.   The code for a 6.8-ohm, 5% resistor is blue, gray, gold, and gold.

d.   Find the maximum and minimum values for a resistor with a color code of orange, orange, blue, and silver.

Orange	Orange	Blue	Silver
3	3	$\times 10^6$	$\pm 10\%$

The average value = 33×10^6 ohms = 33,000,000 ohms = 33 megohms (MΩ)

Maximum value = 33 MΩ + 10% = 33 MΩ + 3.3 MΩ = 36.3 MΩ

Minimum value = 33 MΩ – 10% = 33 MΩ – 3.3 MΩ = 29.7 MΩ

PRACTICAL PROBLEMS

Find the color codes for these resistors.

1.   3.3 kΩ, 5%   _____   _____   _____   _____

2.   820 kΩ, 5%   _____   _____   _____   _____

3.   1.5 MΩ, 10%   _____   _____   _____   _____

4.   10 Ω, 5%   _____   _____   _____   _____

5.   47 kΩ, 10%   _____   _____   _____   _____

Find the values of these resistors.

6.   orange, blue, red, gold   _____

7.   brown, red, blue, silver   _____

8. yellow, orange, black, gold   _____

9. red, yellow, green, gold   _____

10. brown, gray, yellow, silver   _____

11. A 5-kilohm resistor has a 5% tolerance. What is the maximum value the
    resistor can have and still be within the tolerance?   _____

12. The resistor in this diagram has three bands.

    a. Find the minimum resistance.   a. _____

    b. Find the maximum resistance.   b. _____

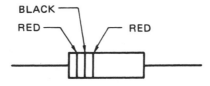

13. An inductor has a value of 120 millihenries and a 20% tolerance.

    a. Find the minimum value.   a. _____

    b. Find the maximum value.   b. _____

14. Find the minimum value of a 25-microfarad capacitor with a 5% tolerance.   _____

15. The values of resistors connected in series are added to form total resistance
    (R_t). Each resistor in this circuit has a 10% tolerance.

    Note: $R_t = R_1 + R_2 + R_3 + R_4$

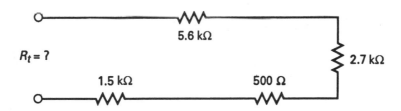

    a. Find the minimum total resistance.   a. _____

    b. Find the maximum total resistance.   b. _____

16. The output of a power supply is specified at 12 volts ±1%.

    a. Find the lower permissible limit.   a. _____

    b. Find the upper permissible limit.   b. _____

17. A 15-microfarad capacitor has a −10%, +80% tolerance.

    a. Find the lower limit.   a. _____

    b. Find the upper limit.   b. _____

18. A resistor has a color code of green, black, orange, and gold. Another resistor is coded yellow, violet, orange, and gold. The tolerances of these resistors overlap by how many ohms?   _____

19. The frequency output of an oscillator is 850 kHz ±3%.

    a. Find the lower limit.   a. _____

    b. Find the upper limit.   b. _____

20. A potentiometer's resistance can be varied from 0 to 10 kilohms. Use this circuit to find these values of the resistance at the point indicated by the formula.

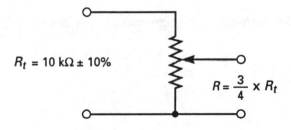

$R_t = 10 \text{ k}\Omega \pm 10\%$

$$R = \frac{3}{4} \times R_t$$

    a. Find the minimum tolerance of the resistance.   a. _____

    b. Find the maximum tolerance of the resistance.   b. _____

AC Circuits

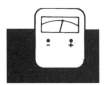

Unit 35   RIGHT TRIANGLES

BASIC PRINCIPLES OF RIGHT TRIANGLES

The *Pythagorean theorem* was developed by a Greek mathematician named Pythagoras almost three thousand years ago. It is a basic law that describes the relationship of the sides of a *right triangle* (a triangle that contains a right, or 90°, angle). The Pythagorean theorem states that the **sum of the squares of the sides of a right triangle is equal to the square of the hypotenuse.** The hypotenuse is the longest side of a right triangle and is located opposite the right or 90° angle.

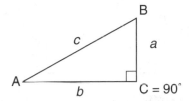

As a general rule, uppercase letters such as *A, B,* and *C* are used to denote an angle. Lowercase letters such as *a, b,* and *c* are used to denote a side. Using the letter *c* to represent the length of the hypotenuse and the letters *a* and *b* to represent the two sides, formulas can be derived to find the length of any side when two others are known using the formula:

$$a^2 + b^2 = c^2$$

The formula for the theorem can be rearranged to give the following equations:

$$c^2 = a^2 + b^2 \qquad c = \sqrt{a^2 + b^2}$$

$$a^2 = c^2 - b^2 \qquad a = \sqrt{c^2 - b^2}$$

$$b^2 = c^2 - a^2 \qquad b = \sqrt{c^2 - a^2}$$

Example: The hypotenuse of a right triangle is 42 feet long. The base is 28 feet in length. Calculate the altitude of the triangle.

$$a = \sqrt{c^2 - b^2} = \sqrt{(42)^2 - (28)^2} = \sqrt{1,764 - 784} = \sqrt{980} = 31.30 \text{ inches}$$

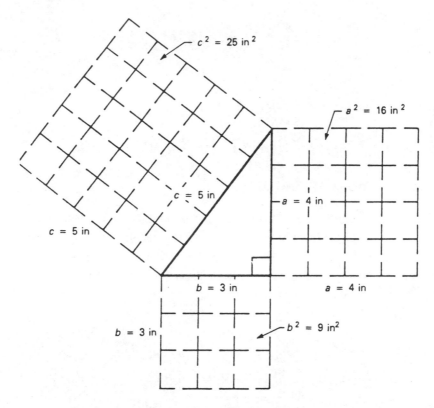

Figure 35–1. The area of the square connected to the base plus the area of the square connected to the altitude is equal to the area of the square connected to the hypotenuse.

The triangle shown in Figure 35–1 is a right triangle with a base 3 inches long, altitude 4 inches long, and hypotenuse 5 inches long. Attached to each side is a square. The base square is $3 \times 3 = 9$ *square* inches. The square using the altitude as one of its sides is $4 \times 4 = 16$ *square* inches. The square on the hypotenuse $5 \times 5 = 25$ *square* inches. Interestingly, the sum of the base and altitude areas ($9 \text{ in}^2 + 16 \text{ in}^2$) equals the square of the hypotenuse (25 in^2). This can be written $c^2 = a^2 + b^2$.

PRACTICAL PROBLEMS

Solve for the unknown value in each of these problems. Round off the answers to two decimal places.

1. $a = 6$, $b = 8$, find c.   _____

2. $a = 30$, $c = 50$, find b.   _____

3. $c = 18$, $b = 12$, find a.   _____

4. $a = 25$, $b = 25$, find c.   _____

5. $a = 4.5$, $c = 10$, find b.   _____

6. $c = 80$, $b = 28$, find a.   _____

7. $a = 100$, $b = 75$, find c.   _____

8. $a = 9\frac{3}{4}$, $c = 15$, find b.   _____

9. $c = 47$, $b = 11$, find a.   _____

10. $a = 90$, $b = 120$, find c.   _____

11.   An antenna tower is supported by two wires. Use this diagram to compute the total length of wire needed to support the tower.   _____

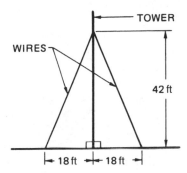

12.   A photocell is mounted in the center of the circuit board in this diagram. How far is the center of the cell from any corner of the board?   _____

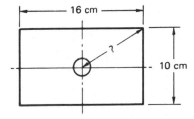

13.   A turntable platter has four equally spaced mounting holes. Use this drawing to determine the distance between holes A and B.   _____

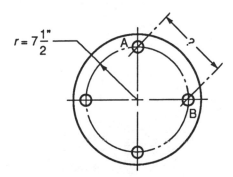

14. A 25-ft ladder is used to install a cable television hookup on the side of a house. The ladder base is 6 feet from the house. How far up the wall does the ladder reach?

_____

15. A single piece of metal is shaped to form an audio console. Use this diagram to determine the total perimeter length of the original piece.

_____

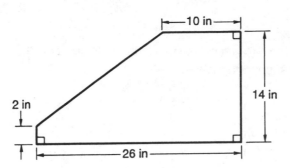

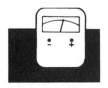

Unit 36   TRIGONOMETRIC FUNCTIONS

BASIC PRINCIPLES OF TRIGONOMETRIC FUNCTIONS

A *triangle* is a closed, three-sided figure. Every triangle, no matter what its shape, has six parts: three sides and three angles. The triangle shown in Figure 36–1 is a *right* triangle, so named because one of its angles (angle C or ∠C) is a right angle or a 90° angle. The right angle is indicated by a small box in a diagram of a right triangle. The horizontal line *b* is the base and the vertical line *a* meeting the base is the *altitude.* The sloping line *c* joining the altitude and base is the *hypotenuse.*

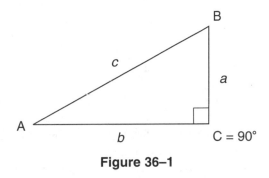

Figure 36–1

Trigonometry is the study of the measures of the sides and angles of triangles. There are three angles in a triangle. The sum of these angles will always be 180°. In the case of the right triangle, since one of the angles is 90°, then the sum of the two remaining angles must also equal 90°.

The angles of a right triangle are determined by the length of its sides. These sides are called the hypotenuse, the opposite side, and the adjacent side. The hypotenuse is always the longest side and is always located opposite the right angle. The side that is opposite or adjacent to the other two angles is determined by the specific angle it is referenced to. The opposite side can be found by bisecting the particular angle of reference. If the bisect line is extended, it will intersect the opposite side. See Figure 36–2 for the relationships described.

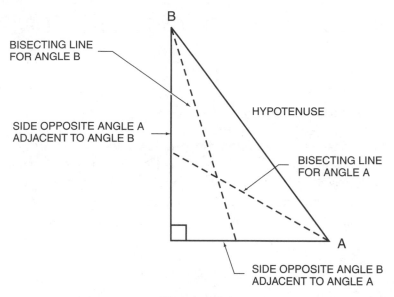

Figure 36–2

The two angles that are less than 90° are formed by the hypotenuse and a side.

There is a special relationship among the sides and angles of a right triangle. Knowing any two sides of a right triangle enables us to calculate the two unknown angles (we know the third is 90°). This is accomplished by determining different ratios of the sides of the triangle with respect to an angle. These ratios are called the *sine, cosine,* and *tangent* of an angle. Table 36–1 presents the values of the sine (sin), cosine (cos), and tangent (tan) for angles ranging from 1° to 90°. Most scientific calculators are able to compute the trigonometric functions for any angle.

Sine of an Angle

The sine of an angle is the ratio of the angle's opposite side to the hypotenuse. This ratio is a constant for a given angle regardless of the size of the right triangle. Referring to Figure 36–1, the sine of ∠A is:

$$\sin A = \frac{\text{opp}}{\text{hyp}} = \frac{a}{c}$$

Example: A right triangle has an altitude of 15 inches and a base of 22 inches. Locate the angle in Table 36–1 that is the closest to the value calculated.

$$\sin A = \frac{\text{opp}}{\text{hyp}} = \frac{15 \text{ in}}{22 \text{ in}} = 0.6818$$

Moving down the sin column in Table 1, we find that angle A is approximately 43° as its sin = 0.6820.

Cosine of an Angle

The cosine of an angle is the ratio of the angle's adjacent side to the hypotenuse. This ratio is a constant for a given angle regardless of the size of the right triangle. Referring to Figure 36–1, the cosine of ∠A is:

$$\cos A = \frac{adj}{hyp} = \frac{b}{c}$$

Example: A right triangle has a base of 12 inches and a hypotenuse of 18 inches. Locate the angle in Table 36–1 that is the closest to the value calculated.

$$\cos A = \frac{adj}{hyp} = \frac{12 \text{ in}}{18 \text{ in}} = 0.6667$$

Moving down the cos column in Table 36–1, we find that angle A is approximately 48° as its cos = 0.6691.

Tangent of an Angle

The tangent of an angle is the ratio of the angle's opposite side to its adjacent side. This ratio is a constant for a given angle regardless of the size of the right triangle. Referring to Figure 1, the tangent of ∠A is:

$$\tan A = \frac{opp}{adj} = \frac{a}{b}$$

Example: A right triangle has an altitude of 20 centimeters and a base of 11.55 centimeters. Locate the angle in Table 36–1 which is the closest to the value calculated.

$$\tan A = \frac{opp}{adj} = \frac{20 \text{ cm}}{11.55 \text{ cm}} = 1.7316$$

Moving down the tan column in Table 36–1, we find that angle A is approximately 60° as its tan = 1.7321.

Table 36–1

			Trigonometric Functions				
Angle	sin	cos	tan	Angle	sin	cos	tan
1°	0.017 5	0.999 8	0.017 5	46°	0.719 3	0.694 7	1.035 5
2°	0.034 9	0.999 4	0.034 9	47°	0.731 4	0.682 0	1.072 4
3°	0.052 3	0.998 6	0.052 4	48°	0.743 1	0.669 1	1.110 6
4°	0.069 8	0.997 6	0.069 9	49°	0.754 7	0.656 1	1.150 4
5°	0.087 2	0.996 2	0.087 5	50°	0.766 0	0.642 8	1.191 8
6°	0.104 5	0.994 5	0.105 1	51°	0.777 1	0.629 3	1.234 9
7°	0.121 9	0.992 5	0.122 8	52°	0.788 0	0.615 7	1.279 9
8°	0.139 2	0.990 3	0.140 5	53°	0.798 6	0.601 8	1.327 0
9°	0.156 4	0.987 7	0.158 4	54°	0.809 0	0.587 8	1.376 4
10°	0.173 6	0.984 8	0.176 3	55°	0.819 2	0.573 6	1.428 1
11°	0.190 8	0.981 6	0.194 4	56°	0.829 0	0.559 2	1.482 6
12°	0.207 3	0.978 1	0.212 6	57°	0.838 7	0.544 6	1.539 9
13°	0.225 0	0.974 4	0.230 9	58°	0.848 0	0.529 9	1.600 3
14°	0.241 9	0.970 3	0.249 3	59°	0.857 2	0.515 0	1.664 3
15°	0.258 8	0.965 9	0.267 9	60°	0.866 0	0.500 0	1.732 1
16°	0.275 6	0.961 3	0.286 7	61°	0.874 6	0.484 8	1.804 0
17°	0.292 4	0.956 3	0.305 7	62°	0.882 9	0.469 5	1.880 7
18°	0.309 0	0.951 1	0.324 9	63°	0.891 0	0.454 0	1.962 6
19°	0.325 6	0.945 5	0.344 3	64°	0.898 8	0.438 4	2.050 3
20°	0.342 0	0.939 7	0.364 0	65°	0.906 3	0.422 6	2.144 5
21°	0.358 4	0.933 6	0.383 9	66°	0.913 5	0.406 7	2.246 0
22°	0.374 6	0.927 2	0.404 0	67°	0.920 5	0.390 7	2.355 9
23°	0.390 7	0.920 5	0.424 5	68°	0.927 2	0.374 6	2.475 1
24°	0.406 7	0.913 5	0.445 2	69°	0.933 6	0.358 4	2.605 1
25°	0.422 6	0.906 3	0.466 3	70°	0.939 7	0.342 0	2.747 5
26°	0.438 4	0.898 8	0.487 7	71°	0.945 5	0.325 6	2.904 2
27°	0.454 0	0.891 0	0.509 5	72°	0.951 1	0.309 0	3.077 7
28°	0.469 5	0.882 9	0.531 7	73°	0.956 3	0.292 4	3.270 9
29°	0.484 8	0.874 6	0.554 3	74°	0.961 3	0.275 6	3.487 4
30°	0.500 0	0.866 0	0.577 4	75°	0.965 9	0.258 8	3.732 1
31°	0.515 0	0.857 2	0.600 9	76°	0.970 3	0.241 9	4.010 8
32°	0.529 9	0.848 0	0.624 9	77°	0.974 4	0.225 0	4.331 5
33°	0.544 6	0.838 7	0.649 4	78°	0.978 1	0.207 9	4.704 6
34°	0.559 2	0.829 0	0.674 5	79°	0.981 6	0.190 8	5.144 6
35°	0.573 6	0.819 2	0.700 2	80°	0.984 8	0.173 6	5.671 3
36°	0.587 8	0.809 0	0.726 5	81°	0.987 7	0.156 4	6.313 8
37°	0.601 8	0.798 6	0.753 6	82°	0.990 3	0.139 2	7.115 4
38°	0.615 7	0.788 0	0.781 3	83°	0.992 5	0.121 9	8.144 3
39°	0.629 3	0.777 1	0.809 8	84°	0.994 5	0.104 5	9.514 4
40°	0.642 8	0.766 0	0.839 1	85°	0.996 2	0.087 2	11.430 1
41°	0.656 1	0.754 7	0.869 3	86°	0.997 6	0.069 8	14.300 7
42°	0.669 1	0.743 1	0.900 4	87°	0.998 6	0.052 3	19.081 1
43°	0.682 0	0.731 4	0.932 5	88°	0.999 4	0.034 9	28.636 3
44°	0.694 7	0.719 3	0.965 7	89°	0.999 8	0.017 5	57.290 0
45°	0.707 1	0.707 1	1.000 0	90°	1.000 0	0.000 0	

FINDING TRIGONOMETRIC FUNCTIONS WITH A CALCULATOR

Scientific calculators are equipped with trigonometric function keys: sine **(SIN)**, cosine **(COS)**, and tangent **(TAN)**. When the calculator is set in the degree **(DEG)** mode, these keys are used to find the sine, cosine, or tangent of different angles. When one of these keys is pressed, it will give the SIN, COS, or TAN of any angle shown on the display or *x* axis.*

Example: With the calculator set in the **DEG** mode, enter 55 on the display and press the **SIN** key.

The number 0.819152044 (depending on the number of digits your calculator can display) will be seen. This is the sine of a 55° angle.

Clear the display and press 55 again. This time press the **COS** key.

This time 0.573576436 is shown on the display. This is the cosine of a 55° angle.

Finding the Angle

The trig function keys can also be used in a reverse operation. They can be used to find an angle when the sine, cosine, or tangent of the angle is known. Different brands of calculators accomplish this in different ways. Some use a second function key **(2nd)** and display the operation as follows:

$$\text{SIN}^{-1} \quad \text{COS}^{-1} \quad \text{TAN}^{-1}$$

Other calculators have an invert **(INV)** or **ARC** key. All of these perform the same function. Assume that it is known that the cosine of an angle is 0.4556, and it is necessary to find the angle that corresponds to this cosine.

Enter .4556

0.4556

Press 2nd, INV, or ARC (depending on the calculator)

Press COS

62.89645407

The number 62.89645407 will be displayed. This is the degree angle that corresponds to a cosine of 0.4556. If the 2nd, INV, or ARC key is not pressed before COS, the number 0.999968385 will be displayed. This is the cosine of a 0.4556° angle.

*On more complex calculators, such as graphing calculators, the trig function keys are pressed before the angle or ratio is entered.

A simple memory aid that can be used to help remember the relationship of the sides of a right triangle to the sine, cosine, and tangent of an angle is: **Oscar Had A Heap Of Apples.** To use this memory aid, write down sin, cos, and tan. The first letter of each word in the saying gives the relationship of the side to the trigonometric function.

$$\text{Sin} \quad \frac{\text{Oscar}}{\text{Had}} \quad \frac{\text{(opposite)}}{\text{(hypotenuse)}}$$

$$\text{Cos} \quad \frac{\text{A}}{\text{Heap}} \quad \frac{\text{(adjacent)}}{\text{(hypotenuse)}}$$

$$\text{Tan} \quad \frac{\text{Of}}{\text{Apples}} \quad \frac{\text{(opposite)}}{\text{(adjacent)}}$$

PRACTICAL PROBLEMS

Use the table of trigonometric functions to locate the specified function for each angle.

1. sin 26° = _____

2. cos 52° = _____

3. tan 2° = _____

4. cos 90° = _____

5. tan 65° = _____

Use the table of trigonometric functions to locate the angle for each function.

6. cos = 0.587 8   ∠ = _____

7. tan = 2.904 2   ∠ = _____

8. sin = 0.121 9   ∠ = _____

9. sin = 0.997 6   ∠ = _____

10. tan = 11.430 1   ∠ = _____

In problems 11–20, select the angle that is closest to the trigonometric function value that is calculated.

Note: Triangles not drawn to scale.

11.   Find ∠A and ∠B in this triangle.

∠A

∠B

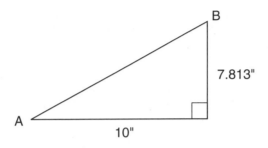

12.   Find ∠A and ∠B in this triangle.

∠A _____

∠B _____

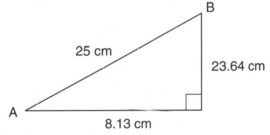

13.   Find ∠A and ∠B in this triangle.

∠A

∠B

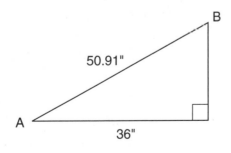

14. The base of a right triangle is 34.64 units in length. The altitude is 20 units in height. What are the three angles within the triangle? (Drawing the triangle may make it easier to solve the problem.)

∠A _____

∠B _____

∠C _____

15. The hypotenuse of a right triangle is 18.63 units in length. The base is 16.45 units. Find angles A and B. (Drawing the triangle may make it easier to solve the problem.)

∠A _____

∠B _____

Find angles A and B in problems 16–20. Note that *c* = hypotenuse, *b* = base, and *a* = altitude. Round angle values to the nearer whole angle.

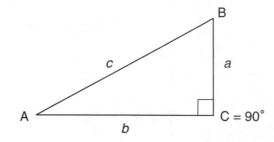

16. *a* = 1.13   *b* = 16.2   ∠A = _____   ∠B = _____

17. *a* = 24.09   *c* = 36   ∠A = _____   ∠B = _____

18. *b* = 250.81   *c* = 487   ∠A = _____   ∠B = _____

19. *a* = 4.23   *b* = 24   ∠A = _____   ∠B = _____

20. *a* = 28.28   *c* = 40   ∠A = _____   ∠B = _____

Unit 37   PLANE VECTORS

BASIC PRINCIPLES OF PLANE VECTORS

A *vector* graphically represents any quantity that has both magnitude and direction. Vectors should not be confused with *scalars*. A scalar represents magnitude only. For example, assume that you ask someone for directions to a certain building and the person says to walk three blocks. This is a scalar because it represents a magnitude only, three blocks. If the person says to walk three blocks east, it is a vector because it contains both magnitude (three blocks) and direction (east).

Rectangular Coordinates

Before beginning the study of vectors, it is helpful to understand rectangular coordinates. The rectangular coordinate system consists of two lines drawn perpendicular to each other (Figure 37–1). The horizontal line is called the *x* axis and the vertical line is called the *y* axis. The point at which the two lines intersect is called the *origin*. Values to the right and above the origin have a positive value, and

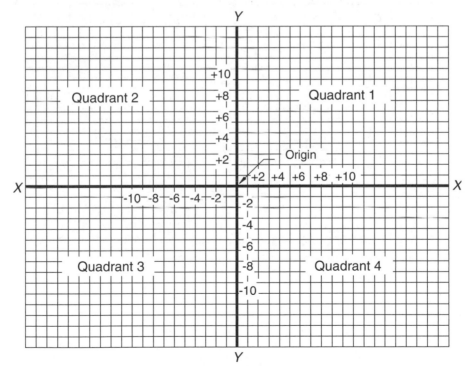

Figure 37–1

values to the left and below the origin have a negative value. The rectangular coordinate system can also be divided into four quadrants as shown in Figure 37–1.

Any point on the plane can be represented by an *X* and *Y* pair of coordinates. The *X* coordinate is called the *abscissa,* and the *Y* coordinate is called the *ordinate.* The origin is given a value of 0,0. In Figure 37–2, the following coordinates have been plotted: (*X*3, *Y*5) (*X*4, *Y*-2) (*X*-2, *Y*6) and (*X*-3, *Y*-4). Although in this example the coordinates are given with both alphabetic and numeric reference, as a general rule coordinates are given using only numeric references. The *X* value is given first, and the *Y* value is give second. Generally *X*3 *Y*5 would be given as (3, 5), and *X*4, *Y*-2 would be given as (4, -2).

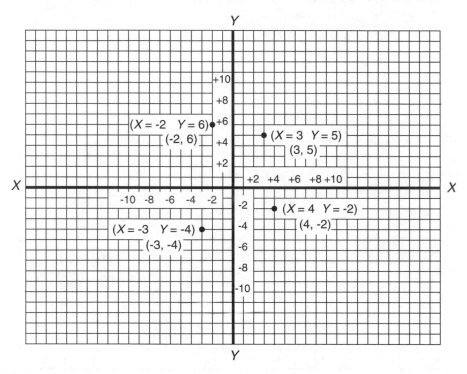

Figure 37–2

When plotting vectors, zero degrees is represented by drawing a horizontal line to the right of the origin. Vectors rotate in a counterclockwise direction around the origin.

Adding Vectors

When determining values in alternating current circuits, it is often necessary to employ vector addition because different types of AC loads cause the current and voltage to become out of phase with each other. These out-of-phase conditions can be represented by vectors. When vectors have the same direction or phase relationship, they can be added together directly (Figure 37–3). A good example of this condition is when two like components are connected together, such as two resistors connected in series. Since both components are resistive, their values would be shown in the same direction, and they would be added algebraically:

$$R_T = R_1 + R_2$$
$$R_T = 70\ \Omega + 140\ \Omega$$
$$R_T = 210\ \Omega$$

where
$$R_1 = 70\ \Omega$$
$$R_2 = 140\ \Omega$$

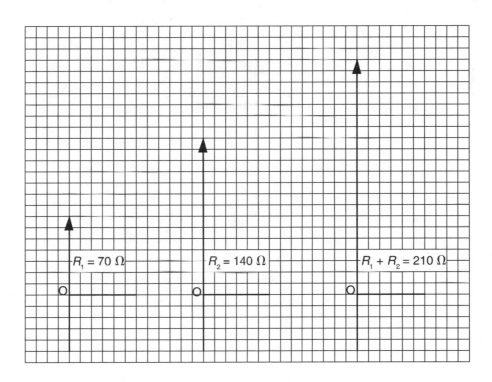

$$R_1 = 70\ \Omega \qquad R_2 = 140\ \Omega$$

Figure 37–3

It is also possible to add two components that are directly opposite in direction or 180° out-of-phase with each other (Figure 37–4). A good example of this would be an inductor and capacitor connected together. Inductive reactance (X_L) and capacitive reactance (X_c) are 180° out-of-phase with each other. When this is the case, it is the same as adding numbers with different signs. You subtract the two quantities. The small quantity is eliminated, and the larger is reduced.

$$X_{Total} = X_L - X_c$$
$$X_{Total} = 240\ \Omega - 160\ \Omega$$
$$X_{Total} = 80\ \Omega$$

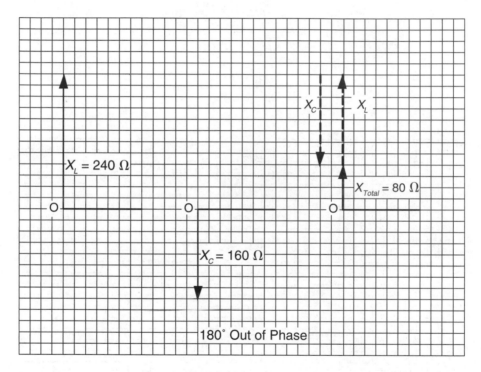

Figure 37–4

Another very common condition in alternating current circuits is for components to be 90° out-of-phase with each other as shown in Figure 37–5. This can be the case when components of capacitance or inductance are connected in the same circuit with resistance. The parallelogram method of vector addition is generally employed to find the resultant of these two values. To use the parallelogram method, draw a second set of vectors that are equal in magnitude and parallel to the first. The resultant, impedance, will be formed from the origin of the two vectors to the intersection point of the parallelogram. Impedance is the total current limiting effect in an alternating current circuit. It is a combination of all current limit values in the circuit such as resistance, inductive reactance, and capacitive reactance. Since the parallelogram in this example contains 90° angles, it forms a rectangle. The resultant forms the hypotenuse of a right triangle. In this condition, the resultant, or impedance, can be found using the Pythagorean theorem.

$$Z = \sqrt{R^2 + X_L^2}$$
$$Z = \sqrt{75^2 + 100^2}$$
$$Z = \sqrt{16{,}525}$$
$$Z = 125 \ \Omega$$

where
$$R = 75 \ \Omega$$
$$X_L = 100 \ \Omega$$

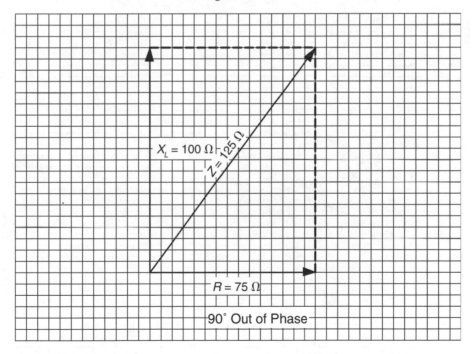

Figure 37–5

The parallelogram method can also be used to find the resultant of vectors when the angle is other than 90°. The *law of cosines* is generally employed to determine the resultant. Assume that one vector (V_1) has a magnitude of 150 and an angle of 0°, and vector 2 (V_2) has a magnitude of 110 and an angle of 55° (Figure 37–6). The law of cosines formula used to determine the resultant of two vectors with an angle less than 90° is shown as follows:

$$OP = \sqrt{V_1^2 + V_2^2 + 2\,(V_1)(V_2)(\cos\angle)}$$
$$OP = \sqrt{150^2 + 110^2 + 2(150)(110)(\cos 55°)}$$
$$OP = \sqrt{22,500 + 12,100 + 2(16,500)\,(0.5736)}$$
$$OP = \sqrt{34,600 + 2(9,464.4)}$$
$$OP = \sqrt{34,600 + 18,928.8}$$
$$OP = \sqrt{53,528.8}$$
$$OP = 231.363$$

where

$$OP = \text{resultant}$$
$$V_1 = 150$$
$$V_2 = 110$$
$$\angle = 55°$$

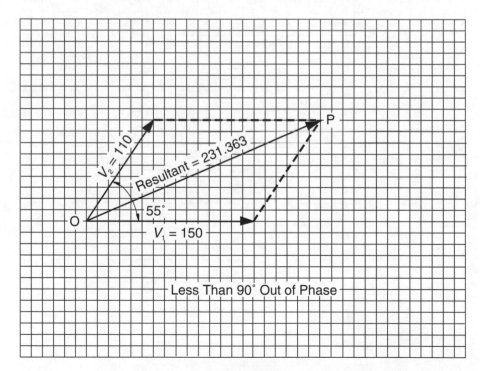

Figure 37–6

A different formula is used for vectors with angles greater than 90° (Figure 37–7). In this example, vector V_1 has a magnitude of 200 and vector V_2 has a magnitude of 80. Vector V_2 is in a direction 101° from vector V_1. The formula shown here is employed to determine the resultant:

$$OP = \sqrt{V_1^2 + V_2^2 - 2\,(V_1)(V_2)(\cos[180 - \angle])}$$
$$OP = \sqrt{200^2 + 80^2 - 2(200)(80)(\cos[180 - 101])}$$
$$OP = \sqrt{46{,}400 - 2(200)(80)(\cos 79)}$$
$$OP = \sqrt{46{,}400 - 2(200)(80)(0.1908)}$$
$$OP = \sqrt{46{,}400 - 2(3{,}052.8)}$$
$$OP = \sqrt{46{,}400 - 6{,}105.6}$$
$$OP = \sqrt{40{,}294.4}$$
$$OP = 200.735$$

where
$$OP = \text{resultant}$$
$$V_1 = 200$$
$$V_2 = 80$$
$$\angle = 101°$$

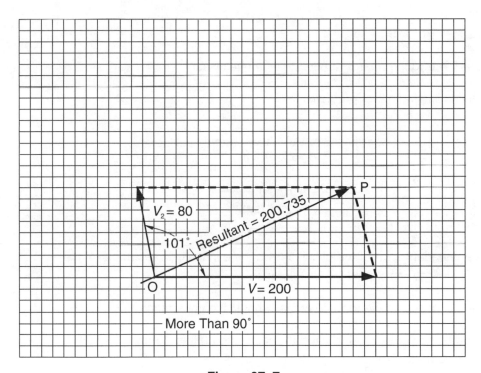

Figure 37–7

Study the electrical impedance diagram in Figure 37–8.

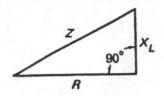

$$Z^2 = R^2 + X_L^2$$
$$Z = \sqrt{R^2 + X_L^2}$$
$$X_L = \sqrt{Z^2 - R^2}$$
$$R = \sqrt{Z^2 - X_L^2}$$

where Z =   impedance

X_L =   inductive reactance

R =   resistance

Figure 37–8

Study the voltage diagram in Figure 37–9.

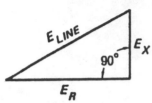

$$E_{LINE} = \sqrt{E_R^2 + E_X^2}$$
$$E_R = \sqrt{E_{LINE}^2 - E_X^2}$$

$$E_X = \sqrt{E_{LINE}^2 - E_R^2}$$

where E_{LINE} =   line voltage

E_R =   voltage across the resistance

E_X =   voltage of the reactance

Figure 37–9

PRACTICAL PROBLEMS

Express the answer to the nearer hundredth when necessary.

Note: Use this illustration for problems 1–3.

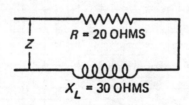

1.   Find the impedance (Z) of the circuit shown if the inductive reactance (X_L) is 30 ohms and the resistance (R) is 20 ohms.   _____

2.   If the resistance is changed to 40 ohms, what is the impedance?   _____

3.   The impedance (Z) is 20 ohms and the resistance (R) is 10 ohms. What is the inductive reactance (X_L)?   _____

4. What is the value of the resistance (*R*) in the series circuit shown?   _____

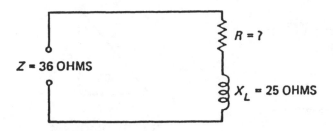

Note: Use this diagram for problems 5–7.

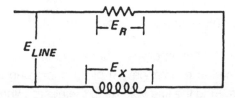

5. Resistance voltage (*E$_R$*) in the voltage diagram is 14 volts, and reactance voltage (*E$_X$*) is 18 volts. What is the line voltage (*E$_{LINE}$*)?   _____

6. The line voltage (*E$_{LINE}$*) is 120 volts, and the voltage of the reactance (*E$_X$*) is 60 volts. What is the voltage across the resistance (*E$_R$*)?   _____

7. Reactance voltage (*E$_X$*) is 150 volts, and resistance voltage (*E$_R$*) is 200 volts. Find the voltage across the line (*E$_{LINE}$*).   _____

8. In the circuit shown, *E$_c$* is 135 volts and *E$_b$* is 175 volts. Find *E$_a$*.   _____

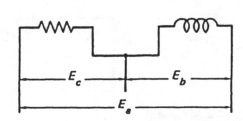

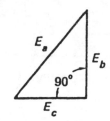

VOLTAGE DIAGRAM

Note: Use this illustration for problems 9–10.

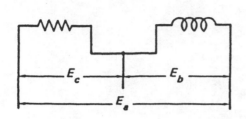

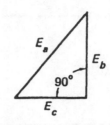

VOLTAGE DIAGRAM

9.   This diagram represents a 3-wire, 2-phase voltage source. If E_b is 130 volts and E_a is 182 volts, find E_c.   _____

10.   Determine the voltage of E_a if E_b is 105 volts and E_c is 120 volts.   _____

11.   A motor draws 3.8 amperes of current when operated from a 60-hertz power source. The motor winding has a series resistance (R) of 30 ohms and an inductive reactance (X_L) of 10 ohms. Find the line voltage to the nearer whole volt. (Hint: First find impedance $Z = \sqrt{R^2 + X_L^2}$.)   _____

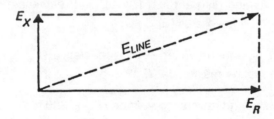

12.   A transformer secondary circuit is shown. Find I_s. Express the answer to the nearer tenth. (Hint: First apply Ohm's law in form $I = \dfrac{E}{R}$ to find I_R and I_L.)   _____

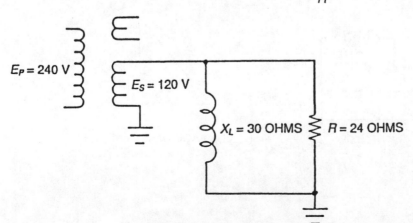

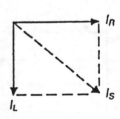

13. A 20-ohm resistor and a capacitor of 15 ohms capacitive reactance at 60 hertz are connected in parallel across a 60-volt, 60-hertz source. Find I_{LINE} to the nearer tenth. (Hint: First apply Ohm's law.)

14. Two voltages (E_1 and E_2) are applied to two resistors. E_1 is 40 volts at 0° reference. E_2 is 60 volts and leads E_1 by 40°. Find E_R to the nearer tenth.

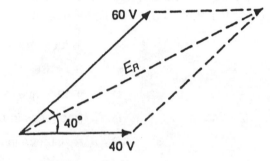

15. Two voltages (E_1 and E_2) are applied to a resistor. E_1 is 60 volts at 0° reference and E_2 is 90 volts and leads E_1 by 104°. Find E_R to the nearer tenth.

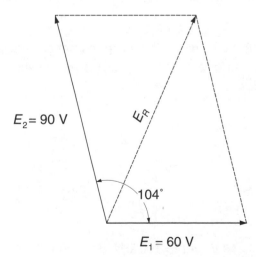

UNIT 38   INDUCTIVE CIRCUITS AND RL CIRCUITS

BASIC PRINCIPLES OF INDUCTIVE CIRCUITS AND RL CIRCUITS

Alternating current (AC) is a current that reverses direction at a certain frequency, which is measured in hertz (Hz). Like direct current, when AC is applied to a circuit, the amount of current that moves through the circuit is dependent on the amount of opposition. In direct current circuits, which is what was considered in previous units, this opposition is resistance. In alternating current circuits, this opposition is more complex.

The opposition to alternating current flow caused by an inductor or coil (Figure 38–1) is *inductive reactance* (X_L). The unit of measurement is the ohm, and the basic formula is given in Figure 38–1.

Figure 38–1

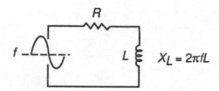

$X_L = 2\pi f L$

X_L = inductive reactance in ohms

π = 3.1416

L = inductance in henries

f = frequency in hertz

R = resistance in ohms

If the resistor in Figure 38–1 was replaced with a short (no resistance), the only opposition to current flow would be the inductive reactance. If the coil was replaced with a short, the only opposition would be the resistance. The combined opposition due to the coil and the resistor is termed *impedance* (Z). The unit of measurement for impedance is the ohm.

Because of the unique qualities of the coil, the relationship among resistance, inductive reactance, and impedance is the same as that of the sides of a right triangle. This relationship and the formula for impedance are presented in Figure 38–2.

Figure 38–2

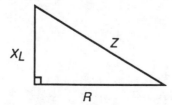

$Z^2 = R^2 + X_L^2$

Z = impedance in ohms

R = resistance in ohms

X_L = inductive reactance in ohms

Example: Given $R = 5 \cdot k\Omega$, $L = 5$ mH, $E = 10$ V, and $f = 100$ kHz, determine the current flowing through the circuit.

Note: When voltage (E_s) is applied to a series resistive-inductive circuit, a current exists. The current (I) is found by the formula $I = \dfrac{E_s}{Z}$, which is another form of Ohm's law.

$$X_L = 2\pi f L$$

$$X_L = 2 \times 3.1416 \times (100 \times 10^3 \text{ Hz}) \times (5 \times 10^{-3} \text{ H}) = 3.14 \text{ k}\Omega$$

$$Z = \sqrt{(5)^2 + (3.14)^2}$$

$$= 5.90 \text{ k}\Omega$$

$$I = \frac{E_s}{Z} = \frac{10\text{V}}{5.9 \text{ k}\Omega} = 1.69 \text{ mA}$$

PRACTICAL PROBLEMS

Note: Use $X_L = 2\pi f L$ for problems 1–2.

1. What reactance does an 8-henry coil have at 1,000 hertz? Express the answer to the nearer hundredth.   _____

2. A 25-millihenry coil is in a circuit operating at 40 kilohertz. Compute the inductive reactance of the coil.   _____

3. What frequency is being applied to the 120-millihenry coil in this diagram?   _____

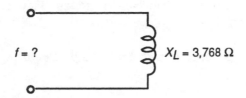

$f = ?$ $X_L = 3,768 \ \Omega$

4. A coil has a 7.5-kilohm reactance to a 2,000-hertz signal. Find the value of the inductor to the nearer millihenry.   _____

5. What size inductor has a reactance of 150 ohms in a 60-hertz circuit?   _____

6. Find the value to the nearest hundredth of a millihenry of the inductor in this circuit. _____

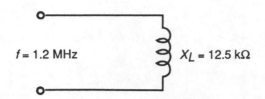

$f = 1.2$ MHz $X_L = 12.5$ kΩ

7. A coil has a reactance of 15 kilohms and is in series with a 12-kilohm resistor. Find the circuit impedance. (Express the answer to the nearer tenth.) _____

   Note: $Z = \sqrt{R^2 + X_L^2}$

8. Find the circuit impedance in this RL circuit. _____

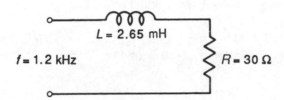

$L = 2.65$ mH

$f = 1.2$ kHz $R = 30\ \Omega$

9. An 80-millihenry coil and a 5-kilohm resistor are in series. Find circuit impedance if a frequency of 4,000 hertz is applied. _____

10. A series RL circuit has a total impedance of 8 kilohms. If the value of X_L is 4 kilohms, find the value of R. _____

    Note: $R = \sqrt{Z^2 - X_L^2}$

11. Impedance of a resistive-inductive circuit is 35 kilohms. Find the inductive reactance if $R = 28$ kilohms. _____

    Note: $X_L = \sqrt{Z^2 - R^2}$

12. Use this circuit to find the approximate value of the inductor. (Hint: Find X_L first.) _____

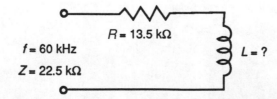

$R = 13.5$ kΩ

$f = 60$ kHz $L = ?$

$Z = 22.5$ kΩ

13. Find the circuit current in this schematic. (Hint: Find Z first.) _____

   Note: $I = \dfrac{E_s}{Z}$

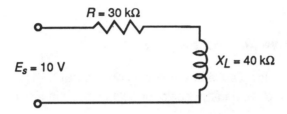

14. Use this circuit to compute the indicated values.

   a. X_L a. _____

   b. Z b. _____

   c. I c. _____

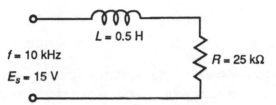

15. Find the source voltage in this RL circuit. _____

   Note: $E_s = IZ$

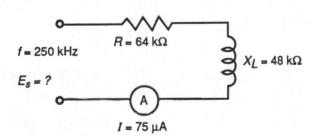

16. A resistor and inductor are connected in series to an 18-volt source operating at a frequency of 2.5 kilohertz. The resistor has a value of 470 ohms, and a current of 26.5 milliamperes flows through it. If the frequency is increased to 3 kilohertz, how much current will flow through the resistor? _____

Unit 39 CAPACITIVE CIRCUITS AND RC CIRCUITS

BASIC PRINCIPLES OF CAPACITIVE AND RC CIRCUITS

The opposition to alternating current flow caused by a capacitor (Figure 39–1) is capacitive reactance (X_c). The unit of measurement is the ohm, and the basic formula is given in Figure 39–1.

Figure 39–1

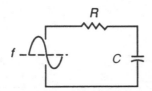

$$X_C = \frac{1}{2\pi fC}$$

X_c = capacitive reactance in ohms

π = 3.1416

C = capacitance in farads

f = frequency in hertz

R = resistance in ohms

If the resistor in Figure 39–1 was replaced with a short (no resistance), the only opposition to current flow would be the capacitive reactance. If the capacitor was replaced with a short, the only opposition would be the resistance. The combined opposition due to the capacitor and the resistor is termed *impedance* (Z). The unit of measurement for impedance is the ohm.

Because of the unique qualities of the capacitor, the relationship among resistance, inductive reactance, and impedance is the same as that of the sides of a right triangle. This relationship and the formula for impedance are presented in Figure 39–2.

Figure 39–2

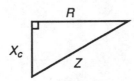

$Z^2 = X_c^2 + R^2$

Z = impedance in ohms

X_c = capacitive reactance in ohms

R = resistance in ohms

Example: Given $R = 500\ \Omega$, $C = 0.1\ \mu F$, $E_s = 10$ V, and $f = 5$ kHz, determine the current flowing through the circuit.

Note: When voltage (E_s) is applied to a series resistive-capacitive circuit, a current exists. The current (I) is found by the formula $I = \dfrac{E_s}{Z}$.

$$X_c = \frac{1}{2 \times 3.1416 \times (5 \times 10^3) \times (0.1 \times 10^{-6})} = 318.47 \; \Omega$$

$$Z = \sqrt{(500)^2 + (318.47)^2} = 592.81 \; \Omega$$

$$I = \frac{E_s}{Z} = \frac{10V}{592.81} = 16.87 \; mA$$

PRACTICAL PROBLEMS

1. Find the capacitive reactance of a 1-microfarad capacitor in a 5-kilohertz circuit.

   Note: $X_C = \dfrac{1}{2\pi fC}$ _____

2. A 0.02-microfarad capacitor is placed in a circuit operating at a frequency of 60 hertz. Find X_C.   _____

3. Find the capacitive reactance in this circuit.   _____

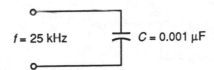

$f = 25 \; kHz$ $C = 0.001 \; \mu F$

4. A 0.5-microfarad capacitor has a reactance of 250 ohms. What frequency is being applied to the capacitor?   _____

   Note: $f = \dfrac{1}{2\pi CX_C}$

5. A circuit operates at a frequency of 455 kilohertz. Approximately what size of capacitor has a reactance of 10 ohms if placed in the circuit?   _____

   Note: $C = \dfrac{1}{2\pi fX_C}$

6.   Find the applied frequency in this circuit.   _____

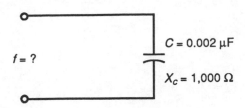

$C = 0.002\ \mu F$

$f = ?$

$X_c = 1,000\ \Omega$

7.   A capacitor has a reactance of 37.5 ohms and is in series with a 50-ohm resistor. Find the circuit impedance.   _____

Note: $Z = \sqrt{R^2 + X_c{}^2}$

8.   Use this circuit to find the indicated values.

a. X_c a. _____

b. Z b. _____

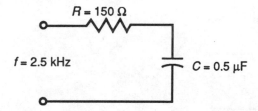

$R = 150\ \Omega$

$f = 2.5\ kHz$

$C = 0.5\ \mu F$

9.   A circuit operating at 75 kilohertz contains a 0.004-microfarad capacitor and a 1,000-ohm resistor in series. Find the indicated values.

a. X_c a. _____

b. Z b. _____

10.   Find the circuit impedance and applied frequency in this circuit as indicated.

a. Z a. _____

b. f b. _____

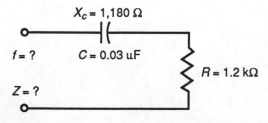

$X_c = 1,180\ \Omega$

$f = ?$ $C = 0.03\ uF$

$R = 1.2\ k\Omega$

$Z = ?$

11. A 0.06-microfarad capacitor has a reactance of 12 kilohms in a series RC
     circuit. The total circuit impedance is 20 kilohms. Find the resistance value.   _____

     Note: $R = \sqrt{Z^2 - X_C^2}$

12. Total impedance in a series RC circuit operating at 60 hertz is 55 kilohms.
     The series resistor has a value of 33 kilohms. Find the capacitive reactance
     and the value of the capacitor as indicated.

     a. X_C a. _____

     b. C b. _____

13. Use the schematic of this RC circuit to find current.   _____

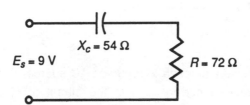

$X_C = 54\ \Omega$

$E_s = 9\ V$ $R = 72\ \Omega$

14. Calculate the indicated values in this circuit.

     a. X_C a. _____

     b. Z b. _____

     c. I c. _____

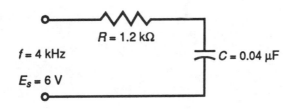

$R = 1.2\ k\Omega$

$f = 4\ kHz$

$C = 0.04\ \mu F$

$E_s = 6\ V$

15. Calculate these values in this series RC circuit.

a. X_C

b. Z

c. E_s

a. _____

b. _____

c. _____

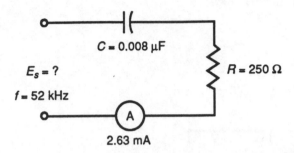

$C = 0.008 \ \mu F$

$E_s = ?$

$f = 52 \ kHz$

$R = 250 \ \Omega$

A

2.63 mA

16. A resistor and capacitor are connected in series to an 8-volt AC source. The frequency is 1 megahertz. There is a current flow in the circuit of 5.17 milliamperes, and the resistor has a resistance of 1,000 ohms. A change of frequency causes the current to increase to 7.2 milliamperes. If the voltage remains constant, what is the new frequency?

_____

Unit 40   RLC CIRCUITS

BASIC PRINCIPLES OF RLC SERIES CIRCUITS

The total opposition to current flow in an alternating current circuit can be a combination of resistance, inductive reactance, and/or capacitive reactance. The total opposition is called impedance. When circuits contain elements of resistance, inductive reactance, and capacitive reactance connected in series, vector addition can be used with the ohmic values of these different components to find the total opposition to current flow.

In the circuit shown in Figure 40–1, a resistor, inductor, and capacitor are connect in series to a 60-hertz AC source. The resistor has a resistance of 40 ohms, the inductor has an inductive reactance of 50 ohms, and the capacitor has a capacitive reactance of 20 ohms. A vector diagram, Figure 40–2, can be used to illustrate the relationship to these three ohmic components. Resistance is drawn on the zero degree axis, inductive reactance is shown to lag the resistance by 90°, and capacitive reactance is shown to lead the resistance by 90°. Since inductive reactance and capacitive reactance are 180° out of phase with each other, the smaller value is subtracted from the larger. This results in the elimination of the smaller value and a reduction of the larger value. The formula shown here can be used to determine the impedance of an RLC series circuit:

$$Z = \sqrt{R^2 + (X_L - X_C)^2}$$
$$Z = \sqrt{40^2 + (50 - 20)^2}$$
$$Z = \sqrt{40^2 + 30^2}$$
$$Z = \sqrt{2,500}$$
$$Z = 50 \ \Omega$$

$R = 40 \ \Omega$

60 Hz

$X_C = 20 \ \Omega$

$X_L = 50 \ \Omega$

Figure 40–1

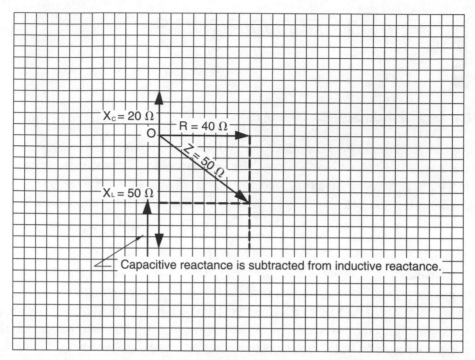

Figure 40–2

In a series circuit, the current is the same throughout the entire circuit. Since voltage and current are in phase with each other in a pure resistive circuit, the voltage drop across the resistor will be in phase with the current. In a pure inductive circuit, the current lags the voltage by 90°. Since the current flow through the inductor and resistor are the same, the voltage drop across the inductor will lead the voltage drop across the resistor by 90°. The voltage dropped across the capacitor will lag the voltage drop across the resistor by 90°. The formula shown here can be used to determine the total voltage applied to the circuit when the voltage drops of each component is known:

$$E_T = \sqrt{E_R{}^2 + (E_L - E_C)^2}$$

where E_R = voltage drop across the resistor

             E_L = voltage drop across the inductor

             E_C = voltage drop across the capacitor

Other formulas for determining values in an RLC series circuit are shown in Figure 40–3.

RLC SERIES CIRCUITS

$$I_T = I_R = I_L = I_C \qquad I_T = \frac{VA}{E_T} \qquad\qquad I_T = \frac{E_T}{Z} \qquad I_T = \sqrt{\frac{VA}{Z}}$$

$$Z = \frac{E_T}{I_T} \qquad\qquad Z = \frac{R}{PF} \qquad\qquad VA = E_T \times I_T \qquad VA = \frac{P}{PF}$$

$$Z = \frac{VA}{I_T^2} \qquad\qquad E_T = \frac{VA}{I_T} \qquad\qquad VA = I_T^2 \times Z \qquad VA = \frac{E_T^2}{Z}$$

$$E_T = I_T \times Z \qquad\qquad E_T = \frac{E_R}{PF} \qquad\qquad VA = \sqrt{P^2 + (\,Vars_L - Vars_C\,)^2}$$

$$PF = \frac{R}{Z} \qquad P = E_R \times I_R \qquad P = VA \times PF \qquad E_R = I_R \times R$$

$$PF = \frac{P}{VA} \qquad P = \sqrt{VA^2 - (\,Vars_L - Vars_C\,)^2} \qquad E_R = \frac{P}{I_R}$$

$$PF = \frac{E_R}{E_T} \qquad P = \frac{E_R^2}{R} \qquad E_R = \sqrt{P \times R} \qquad E_R = \sqrt{E_T^2 - (\,E_L - E_C\,)^2}$$

$$PF = Cos \angle\Theta \qquad P = I_R^2 \times R \qquad E_R = E_T \times PF$$

$$I_R = I_T - I_C = I_L \qquad R = \sqrt{Z^2 - (\,X_L - X_C\,)^2} \qquad E_C = I_C \times X_C$$

$$I_R = \frac{E_R}{R} \qquad\qquad R = \frac{E_R}{I_R} \qquad R = Z \times PF \qquad E_C = \sqrt{Vars_C \times X_C}$$

$$I_R = \frac{P}{E_R} \qquad\qquad R = \frac{E_R^2}{P} \qquad R = \frac{P}{I_R^2} \qquad E_C = \frac{Vars_C}{I_C}$$

$$I_R = \sqrt{\frac{P}{R}} \qquad\qquad X_C = \frac{1}{2 \times \pi \times F \times C} \qquad C = \frac{1}{2 \times \pi \times F \times X_C}$$

$$I_C = I_R = I_T = I_L \qquad X_C = \frac{E_C^2}{Vars_C} \qquad X_C = \frac{Vars_C}{I_C^2} \qquad Vars_C = E_C \times I_C$$

$$I_C = \frac{E_C}{X_C} \qquad\qquad X_C = \frac{E_C}{I_C} \qquad L = \frac{X_L}{2 \times \pi \times F} \qquad X_L = 2 \times \pi \times F \times L$$

$$I_C = \frac{Vars_C}{E_C} \qquad\qquad Vars_C = I_C^2 \times X_C \qquad X_L = \frac{E_L}{I_L} \qquad X_L = \frac{Vars_L}{I_L^2}$$

$$I_C = \sqrt{\frac{Vars_C}{X_C}} \qquad Vars_C = \frac{E_C^2}{X_C} \qquad X_L = \frac{E_L^2}{Vars_L}$$

Figure 40–3

RLC Parallel Circuits

Resistors, inductors, and capacitors can also be connected in parallel as shown in Figure 40–4. As with any type of parallel circuit, the voltage drop across each branch is the same. Since there is a phase angle difference between voltage and current for inductors and capacitors, the total current must be determined by vector addition. Assume that the resistive branch in Figure 40–4 has a current flow of 8 amperes, the inductive branch as a current flow of 12 amperes, and the capacitive branch has a current flow of 6 amperes. The total circuit current can be determined using this formula:

$$I_T = \sqrt{I_R^2 + (I_L - I_C)^2}$$

$$I_T = \sqrt{8^2 + (12 - 6)^2}$$

$$I_T = \sqrt{8^2 + 6^2}$$

$$I_T = \sqrt{64 + 36}$$

$$I_T = \sqrt{100}$$

$$I_T = 10 \text{ A}$$

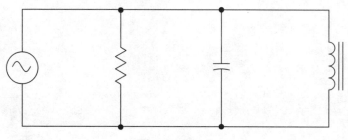

Figure 40–4

A different formula must be used to determine the total impedance in a parallel RLC circuit. Assume that the resistor in Figure 40–4 has a resistance of 120 ohms, the inductor has an inductive reactance of 160 ohms, and the capacitor has a capacitive reactance of 100 ohms. The total impedance can be determined using this formula:

$$Z = \cfrac{1}{\sqrt{\left(\dfrac{1}{R}\right)^2 + \left(\dfrac{1}{X_L} - \dfrac{1}{X_C}\right)^2}}$$

$$Z = \cfrac{1}{\sqrt{\left(\dfrac{1}{120}\right)^2 \left(\dfrac{1}{160} - \dfrac{1}{100}\right)^2}}$$

$$Z = \cfrac{1}{\sqrt{0.000083507}}$$

$$Z = 109.431 \ \Omega$$

To work this problem with a scientific calculator, press the following keys:

$\boxed{1}$ $\boxed{2}$ $\boxed{0}$ $\boxed{1/x}$ $\boxed{x^2}$ $\boxed{+}$ $\boxed{(\,}$ $\boxed{1}$ $\boxed{6}$ $\boxed{0}$ $\boxed{1/x}$ $\boxed{-}$ $\boxed{1}$ $\boxed{0}$ $\boxed{0}$ $\boxed{1/x}$ $\boxed{)}$ $\boxed{x^2}$ $\boxed{=}$ $\boxed{sqr}$ $\boxed{1/x}$

Note: sqr stands for square root

Other formulas for determining values of RLC parallel circuits are shown in Figure 40–5.

RESISTIVE-INDUCTIVE-CAPACITIVE PARALLEL CIRCUITS

$$E_T = \frac{VA}{I_T} \qquad\qquad E_T = I_T \times Z \qquad\qquad E_T = E_R = E_L = E_C$$

$$E_T = \sqrt{VA \times Z} \qquad\qquad Z = \frac{VA}{I_T^2} \qquad\qquad I_T = \frac{VA}{E_T} \qquad I_T = \sqrt{\frac{VA}{Z}}$$

$$Z = \frac{E_T}{I_T} \quad Z = \frac{E_T^2}{VA} \quad Z = R \times PF \qquad I_T = \frac{E_T}{Z} \qquad\qquad I_T = \frac{I_R}{PF}$$

$$VA = E_T \times I_T \qquad\qquad PF = \frac{Z}{R} \qquad\qquad PF = \text{Cos} \angle \Theta \qquad E_L = E_T = E_R = E_C$$

$$VA = I_T^2 \times Z \qquad\qquad PF = \frac{P}{VA} \qquad E_L = I_L \times X_L \qquad E_L = \sqrt{\text{Vars}_L \times X_L}$$

$$VA = \frac{E_T^2}{Z} \qquad\qquad PF = \frac{I_R}{I_T} \qquad E_L = \frac{\text{Vars}_L}{I_L} \qquad L = \frac{X_L}{2 \times \pi \times F}$$

$$VA = \sqrt{P^2 + (\text{Vars}_L - \text{Vars}_C)^2} \quad X_L = \frac{E_L}{I_L} \qquad X_L = 2 \times \pi \times F \times L \quad \text{Vars}_L = \frac{E_L^2}{X_L}$$

$$VA = \frac{P}{PF} \qquad\qquad I_L = \frac{\text{Vars}_L}{E_L} \qquad X_L = \frac{E_L^2}{\text{Vars}_L} \qquad \text{Vars}_L = I_L^2 \times X_L$$

$$I_L = \frac{E_L}{X_L} \qquad\qquad I_L = \sqrt{\frac{\text{Vars}_L}{X_L}} \qquad X_L = \frac{\text{Vars}_L}{I_L^2} \qquad \text{Vars}_L = E_L \times I_L$$

$$E_R = I_R \times R \qquad\qquad I_R = \sqrt{I_T^2 - (I_L - I_C)^2} \qquad I_R = I_T \times PF$$

$$E_R = \sqrt{P \times R} \qquad\qquad I_R = \frac{E_R}{R} \qquad R = \frac{E_R}{I_R} \qquad\qquad R = \cfrac{1}{\sqrt{\left(\frac{1}{Z}\right)^2 - \left(\frac{1}{X_L} - \frac{1}{X_C}\right)^2}}$$

$$E_R = \frac{P}{I_R} \qquad\qquad I_R = \frac{P}{E_R}$$

$$E_R = E_T = E_L = E_C \quad I_R = \sqrt{\frac{P}{R}} \qquad R = \frac{P}{I_R^2} \qquad R = \frac{Z}{PF} \qquad R = \frac{E_R^2}{P}$$

$$P = \sqrt{VA^2 - (\text{Vars}_L - \text{Vars}_C)^2} \qquad P = E_R \times I_R \qquad P = I_R^2 \times R$$

$$E_C = \frac{\text{Vars}_C}{I_C} \qquad\qquad P = VA \times PF \quad P = \frac{E_R^2}{R} \qquad X_C = \frac{1}{2 \times \pi \times F \times C}$$

$$E_C = I_C \times X_C \qquad\qquad I_C = \frac{E_C}{X_C} \qquad X_C = \frac{E_C}{I_C} \qquad \text{Vars}_C = E_C \times I_C$$

$$E_C = E_T = E_R = E_L \quad I_C = \frac{\text{Vars}_C}{E_C} \qquad X_C = \frac{E_C^2}{\text{Vars}_C} \qquad \text{Vars}_C = I_C^2 \times X_C$$

$$E_C = \sqrt{\text{Vars}_C \times X_C} \quad I_C = \sqrt{\frac{\text{Vars}_C}{X_C}} \qquad X_C = \frac{\text{Vars}_C}{I_C^2} \qquad \text{Vars}_C = \frac{E_C^2}{X_C}$$

Figure 40–5

PRACTICAL PROBLEMS

1. A coil has an inductive reactance of 12.5 kilohms. It is in a series RCL circuit with a capacitor that has a capacitive reactance of 8.7 kilohms. Find the total reactance (X).

    _____

2. An RCL series circuit has $R = 15$ kΩ, $X_C = 36.8$ kΩ, and $X_L = 16.8$ kΩ. Calculate the circuit impedance (Z).

    _____

Note: Use this circuit for problems 3–6.

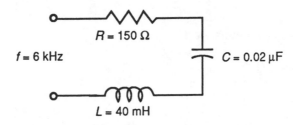

$R = 150\ \Omega$

$f = 6$ kHz

$C = 0.02\ \mu F$

$L = 40$ mH

3. What is the value of the inductive reactance (X_L)?

    _____

4. Find the value of the capacitive reactance (X_C).

    _____

5. Compute circuit reactance (X).

    _____

6. What is the total circuit impedance (Z)?

    _____

Note: Use this circuit for problems 7–12.

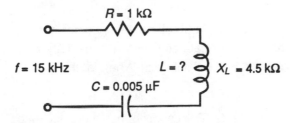

$R = 1$ kΩ

$f = 15$ kHz

$L = ?$ $X_L = 4.5$ kΩ

$C = 0.005\ \mu F$

7. Find the value of the inductor (L).

    _____

8. What is the capacitive reactance (X_C)?

    _____

9. Compute circuit reactance (X).

    _____

10. Find the total impedance (Z).

    _____

11. What is the value of X_L if the frequency is decreased to 10 kilohertz?

    _____

12. Calculate the total impedance if the frequency is set to 10 kilohertz.

    _____

Note: Use this circuit for problems 13–16.

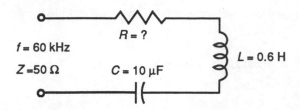

13. Find the value of X_L.   _____

14. Compute the value of the capacitive reactance (X_C).   _____

15. What is the total circuit reactance (X)?   _____

16. Find the value of resistance (R).   _____

Note: Use this circuit for problems 17–20.

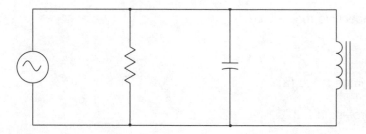

17. Assume that the resistor has a resistance of 300 ohms, the inductor has an inductance of 7.162 millihenries, and the capacitor has a capacitance of 79.577 microfarads. The frequency is 10 kilohertz. What is the circuit impedance?   _____

18. The circuit has a frequency of 250 kilohertz. The capacitor has a capacitance of 636.618 picofarads and a current flow of 24 milliamperes. What is the voltage drop across the resistor?   _____

19. The resistor has a current flow of 8 milliamperes. The inductor has a current flow of 12 milliamperes, and the capacitor has a current flow of 20 milliamperes. What is the total circuit current?   _____

20. The circuit has a total current flow of 9.112 milliamperes. The current through the inductor is 13.333 milliamperes, and the current through the capacitor is 18.824 milliamperes. How much current is flowing through the resistor?   _____

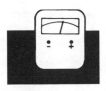

Unit 41 SIMULTANEOUS EQUATIONS

BASIC PRINCIPLES OF SIMULTANEOUS EQUATIONS

Simultaneous equations are generally used to solve complex circuits that contain more than one power source. Problems that are solved using Kirchhoff's law are a good example of these types of circuits. Simultaneous means "at the same time." Simultaneous equations are necessary when an equation contains more than one unknown.

Example: Consider this equation:

$$x + y = 10$$

It would be impossible to determine the values of x and y in this equation. The value of x could be 5, the value of y could be 5. Or x could equal 6 and y equal 4, or x could equal 8 and y could equal 2. As you can see, there is no way to determine the answer.

The equation can be solved, however, if another equation containing the same unknowns can be found. Consider the equation:

$$x - y = 4$$

The unknown values can be determined by adding the two equations together:

$$x + y = 10$$

$$x - y = 4$$

Since only like terms can be added together, the two equations should be arranged so that like terms are grouped. Here x plus x equals $2x$. A $+y$ added to a $-y$ equals zero. For this reason, y will disappear from the equation.

$$2x = 14$$
$$x = \frac{14}{2}$$
$$x = 7$$

Now that the value of x has been determined, it can be substituted in one of the original equations to find the value of y.

$$7 + y = 10$$

$$y = 10 - 7$$

$$y = 3$$

Example: Consider this equation:

$$a + 2b = 20$$
$$\underline{a - b = 2}$$

Hint: In algebra you can do anything to an equation as long as it is done on both sides of the equal sign. When solving simultaneous equations, eliminate one unknown by making them cancel. This is done here by multiplying the lower equation by 2:

$$a + 2b = 20$$
$$\underline{2(a - b = 2)}$$

$$a + 2b = 20$$

$$2a - 2b = 4$$

Minus b multiplied by plus 2 becomes minus $2b$. When plus $2b$ and minus $2b$ are added, the sum is zero. The equation now becomes:

$$3a = 24$$
$$a = \frac{24}{3}$$
$$a = 8$$

Now that one value has been determined, the other can be found by substituting the known value in one of the original formulas:

$$8 + 2b = 20$$

$$2b = 20 - 8$$

$$2b = 12$$

$$b = \frac{12}{2}$$

$$b = 6$$

Each of the two example problems contained two unknown quantities. Two equations are required to solve for two unknown quantities. If an equation contains three unknown quantities, it requires three separate equations to find the unknown values.

Kirchhoff's Current Law

As stated previously, simultaneous equations are often employed in the electronics field to solve problems involving Kirchhoff's law. For that reason, Kirchhoff's law will be covered in this unit.

Kirchhoff's current law basically states that the algebraic sum of currents entering and leaving any particular point must equal zero. Consider the example shown in Figure 41–1. The current flow through resistor R_1 is 4 amperes, and the current flow through resistor R_2 is 6 amperes. Both currents enter point P. The current exiting point P is 10 amperes. Kirchhoff's current law, however, states that the sum of the currents entering and leaving a particular point must equal zero. When using Kirchhoff's law, the current entering a point is considered positive and the current leaving a point is considered negative.

$$4\,A + 6\,A - 10\,A = 0$$

Currents I_1 and I_2 are considered positive because they enter point P, and current I_{1+2} is considered negative because it leaves point P.

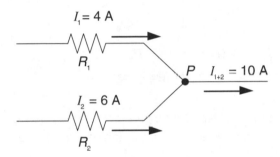

Figure 41–1

A second circuit illustrating Kirchhoff's current law is shown in Figure 41–2. Consider what happens to the current at point B. Two amperes flow into point B from resister R_1. The current then splits so that 1.2 amperes flow to resistors R_4, R_5, and R_6, and 0.8 ampere flows to resistor R_2. The current entering point B is positive, and the currents leaving point B are negative.

$$I_1 - I_2 - I_{4-6} = 0$$

$$2\,A - 0.8\,A - 1.2\,A = 0$$

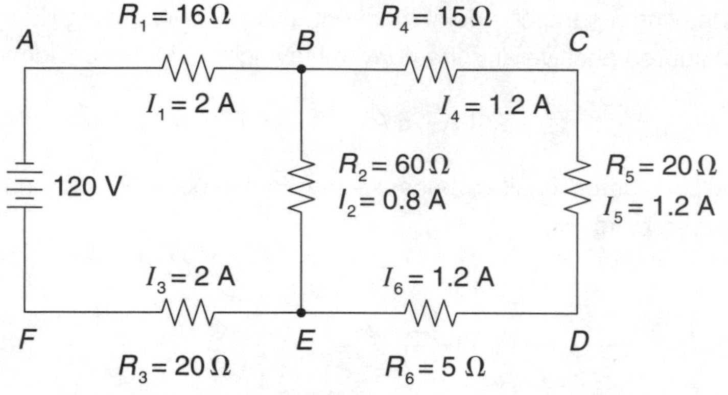

Figure 41–2

Now consider what happens at point E. Here 0.8 ampere enters from resistor R_2, and 1.2 amperes enter from resistors R_4, R_5, and R_6. Two amperes exit point E and flow to resistor R_3.

$$0.8\,A + 1.2\,A - 2\,A = 0$$

Kirchhoff's Voltage Law

Kirchhoff's voltage law states that the sum of the voltage drops around any closed loop must equal zero. Before you can determine the algebraic sum of the voltages, you must determine which end of the element is positive and which is negative. To make this determination, assume a direction of current flow and mark the end of the resistive element where current enters and where current leaves. It will be assumed that current flows from negative to positive. Therefore, the point where current enters the element will be marked negative, and the point where current leaves will be marked positive. Voltage drops and polarity marking have been added to Figure 41–2, and the amended circuit is shown in Figure 41–3. To use Kirchhoff's voltage law, start at some point and add the voltage drops around any closed loop. Be certain to return to the starting point. In the circuit shown, there are actually three separate closed loops to be considered. Loop $ACDF$ contains the voltage drops E_1, E_4, E_5, E_6, E_3 and E_T (the 120-volt source). Loop $ABEF$ contains the voltage drops E_1, E_2, E_3, and E_T. Loop $BCDE$ contains the voltage drops E_4, E_5, E_6, and E_2. The voltage drops for the first loop are as follows:

$$-E_1 - E_4 - E_5 - E_6 - E_3 - E_T = 0$$

$$-32 - 18 - 24 - 5 - 40 + 120 = 0$$

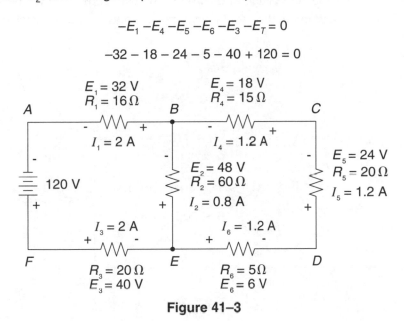

Figure 41–3

The positive or negative sign for each number is determined by the assumed direction of current flow. In this example, it is assumed that current leaves point A and returns to point A. The current leaving point A enters resistor R_1 at the negative end. Therefore, the voltage is considered to be negative (-32 V). The same is true for resistors R_4, R_5, and R_6, and R_3. The current enters the voltage source, E_T, at the positive end. Therefore, E_T is assumed to have a positive value.

For the second loop, it is assumed that current will leave point A and returns to point A through resistors R_1, R_2, R_3, and R_T. The voltage drops are as follows:

$$-32 - 48 - 40 + 120 = 0$$

The third loop assumed that current leaves point B and returns to point B. The current path is through resistors R_4, R_5, R_6, and R_2. The voltage drops are as shown:

$$-18 - 24 - 6 + 48 = 0$$

Applying Kirchhoff's Law

So far, the values of voltage and current have been given to illustrate how Kirchhoff's law is applied to a circuit. In the circuit shown in Figure 41–4, the voltage drops and currents are not known. This circuit contains three resistors and two voltage sources. Although there are three separate loops, only two are needed to find the missing values of voltage and current. The two loops used will be $ABEF$ and $CBED$. For the first loop, it will be assumed that current leaves point A and returns to point A. The equation for this loop is:

$$-E_1 - E_3 + E_{S_1} = 0$$

$$-E_1 - E_3 + 60 = 0$$

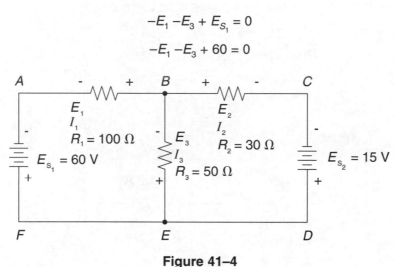

Figure 41–4

The current path for the second loop leaves point C and returns to point C. The equation for the second loop is:

$$-E_2 - E_3 + E_{S_2} = 0$$

$$-E_2 - E_3 + 15 = 0$$

To simplify the two equations, the whole numbers will be moved to the other side of the equal sign. This will be done in the first equation by subtracting 60 from both sides of the equation, and in the second equation 15 will be subtracted from both sides:

$$-E_1 - E_3 = -60$$

$$-E_2 - E_3 = -15$$

The two equations can be further simplified by changing the negative whole numbers into positive whole numbers. This is accomplished by multiplying both equations by negative 1 (−1). This causes all the signed numbers to change. The new equations are:

$$E_1 + E_3 = 60$$

$$E_2 + E_3 = 15$$

Although there are now two equations, they cannot be added because only *like* terms can be added. The two values of E_3 can be added because they are alike, but E_1 and E_2 cannot. This problem can be solved by changing the terms to their Ohm's law equivalents. According to Ohm's law, the voltage drop across any resistive element is equal to the resistance of the resistor and the amount of current flowing through it ($E = I \times R$). The three terms E_1, E_2, and E_3 will be changed to their Ohm's law equivalents.

$$E_1 = I_1 \times R_1 \qquad I_1 \times 100 = 100I_1$$

$$E_2 = I_2 \times R_2 \qquad I_2 \times 30 = 30I_2$$

Although it is true that $E_3 = I_3 \times R_3$, this would produce three unknown terms in the equation. Since Kirchhoff's law states that the sum of the currents entering a point must equal the current leaving a point, I_3 is actually the sum of the current I_1 and I_2. The third voltage equation can be written:

$$E_3 = (I_1 + I_2) \times R_3 \qquad (I_1 + I_2) \times 50 = 50(I_1 + I_2)$$

The two equations can now be written as:

$$100I_1 + 50(I_1 + I_2) = 60$$

$$30I_2 + 50(I_1 + I_2) = 15$$

The parentheses can be removed by multiplying both terms by 50. The equations now become:

$$100I_1 + 50I_1 + 50I_2 = 60$$

$$30I_2 + 50I_1 + 50I_2 = 15$$

Gathering terms is accomplished by adding like terms together. The two equations now become:

$$150I_1 + 50I_2 = 60$$

$$50I_1 = +80I_2 = 15$$

The two equations can now be solved as simultaneous equations. To do this, one of the terms must be eliminated. In this example, the bottom equation will be multiplied by negative 3 (–3).

$$-3(50I_1 + 80I_2 = 15)$$

The product becomes:

$$-150I_1 - 240I_2 = -45$$

The positive $150I_1$ and negative $150I_1$ will now cancel each other. The two I_2 terms can be added:

$$150I_1 + 50I_2 = 60$$

$$\underline{-150I_1 - 240I_2 = -45}$$

$$-190I_2 = 15$$

Dividing both sides of the equations by –190 will produce the answer for I_2:

$$I_2 = -0.0789 \text{ A}$$

The negative answer indicates that the assumed current direction was incorrect. Current actually flows through resistor R_2 in the opposite direction as shown in Figure 41–4.

Now that the value of I_2 has been determined, the value can be substituted in either of the formulas to find the value of I_1:

$$150I_1 + 50(-0.0789) = 60$$

$$150I_1 - 3.945 = 60$$

Now add +3.945 to both side of the equation:

$$150I_1 = 63.945$$

$$I_1 = \frac{63.945}{150}$$

$$I_1 = 0.426 \text{ A}$$

Since the I_3 is actually the sum of current I_1 and I_2:

$$I_3 = I_1 + I_2$$

$$I_3 = 0.426 - 0.0789$$

$$I_3 = 0.347 \text{ A}$$

The voltage drop across each resistor can now be determined using Ohm's law:

$$E_1 = 0.426 \times 100$$

$$E_1 = 42.6 \text{ V}$$

$$E_2 = 0.00789 \times 30$$

$$E_2 = 2.367 \text{ V}$$

$$E_3 = 0.347 \times 50$$

$$E_3 = 17.35 \text{ V}$$

PRACTICAL PROBLEMS

Solve the following equations:

1. $3x + 2y = 77$ $x =$ _____

    $4x - y = 77$ $y =$ _____

2. $2x + 5y = 34$ $x =$ _____

    $4x + 2y = 34.4$ $y =$ _____

3. $3a - 4b = 64$ $a =$ _____

    $2a + 2b = 24$ $b =$ _____

4. $25m + 20n = 165$ $m =$ _____

    $5m + 8n = 51$ $n =$ _____

5. $z + 4j = -195$ $z =$ _____

    $8z - j = -75$ $j =$ _____

6. $12x - 3y = 6.6$ $x =$ _____

    $3x + 5y = 2.8$ $y =$ _____

7. $3w - 2z = 10$

    $2w + 4z = 124$

    $w =$ _____

    $z =$ _____

8. $4c - 6d = -204$

    $c + 2d = 12$

    $c =$ _____

    $d =$ _____

9. Find the unknown values in this circuit.

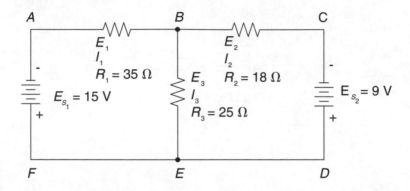

$I_1 =$

$I_2 =$

$I_3 =$

$E_1 =$

$E_2 =$

$E_3 =$

10.   Find the unknown values in this circuit.

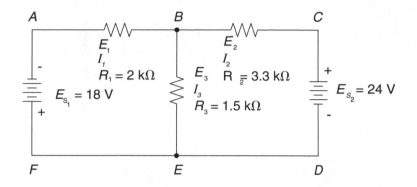

$I_1 =$ _____

$I_2 =$ _____

$I_3 =$ _____

$E_1 =$ _____

$E_2 =$ _____

$E_3 =$ _____

Appendix

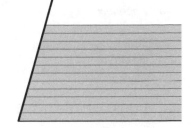

SECTION I

DENOMINATE NUMBERS

Denominate numbers are numbers that include units of measurement. The units of measurement are arranged from the largest unit at the left to the smallest unit at the right.

For example: 6 yd 2 ft 4 in is the correct representation of a denominate number.

All basic operations of arithmetic can be performed on denominate numbers.

I.   EQUIVALENT MEASURES

A measurement can be expressed in an equivalent form, but with different units, through multiplication by a *conversion factor.* For example, the conversion factor that changes feet to inches or inches to feet is

$$12 \text{ in} = 1 \text{ ft}$$

To change a measurement given in inches to one in feet, multiply by the conversion factor $\dfrac{1 \text{ ft}}{12 \text{ in}}$.

To change a measurement given in feet to one in inches, multiply by the conversion factor $\dfrac{12 \text{ in}}{1 \text{ ft}}$.

Example: To express 6 inches in equivalent foot measurement, multiply by $\dfrac{1 \text{ ft}}{12 \text{ in}}$. In the numerator and denominator, divide by a common factor.

$$6 \text{ in} = \frac{\overset{1}{\cancel{6}} \text{ in}}{1} \times \frac{1 \text{ ft}}{\underset{2}{\cancel{12}} \text{ in}} = \frac{1}{2} \text{ ft or } 0.5 \text{ ft}$$

To express 4 feet in equivalent inch measurement, multiply 4 feet by $\dfrac{12 \text{ in}}{1 \text{ ft}}$.

$$4 \text{ ft} = \frac{4 \text{ ft}}{1} \times \frac{12 \text{ in}}{1 \text{ ft}} = \frac{48 \text{ in}}{1} = 48 \text{ in}$$

Per means division, as with a fraction bar. For example, 50 miles per hour can be written $\frac{50 \text{ miles}}{1 \text{ hour}}$.

II.   BASIC OPERATIONS

A.   ADDITION

EXAMPLE: 2 yd 1 ft 5 in + 1 ft 8 in + 5 yd 2 ft

1. Write the denominate numbers in a column with like units in the same column.

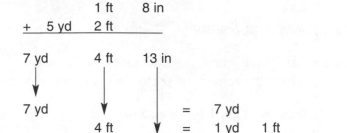

	2 yd	1 ft	5 in
		1 ft	8 in
+	5 yd	2 ft	

2. Add the denominate numbers in each column.

	7 yd	4 ft	13 in

3. Express the answer using the largest possible units.

7 yd = 7 yd
4 ft = 1 yd 1 ft
13 in = + 1 ft 1 in

7 yd   4 ft   13 in = 8 yd 2 ft 1 in

B.   SUBTRACTION

EXAMPLE: 4 yd 3 ft 5 in − 2 yd 1 ft 7 in

1. Write the denominate numbers in columns with like units in the same column.

	4 yd	3 ft	5 in
−	2 yd	1 ft	7 in

2. Starting at the right, examine each column to compare the numbers. If the bottom number is larger, exchange one unit from the column at the left for its equivalent. Combine like units.

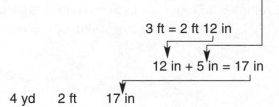

7 in is larger than 5 in

3 ft = 2 ft 12 in

12 in + 5 in = 17 in

3. Subtract the denominate numbers.

	4 yd	2 ft	17 in
−	2 yd	1 ft	7 in
	2 yd	1 ft	10 in

4. Express the answer using the largest possible units.

	2 yd	1 ft	10 in

C.   MULTIPLICATION

—By a constant

EXAMPLE: 1 hr 24 min × 3

1.   Multiply the denominate number
     by the constant.

$$\begin{array}{rr} 1\ hr & 24\ min \\ \times & 3 \\ \hline 3\ hr & 72\ min \end{array}$$

2.   Express the answer using the
     largest possible units.

3 hr   = 3 hr

    72 min   = 1 hr   12 min

3 hr   72 min   = 4 hr   12 min

—By a denominate number expressing linear measurement

EXAMPLE: 9 ft 6 in × 10 ft

1.   Express all denominate numbers
     in the same unit.

$$9\ ft\ 6\ in = 9\frac{1}{2}\ ft$$

2.   Multiply the denominate numbers.
     (This includes the units of measure,
     such as ft × ft = sq ft.)

$$9\frac{1}{2}\ ft \times 10\ ft =$$

$$\frac{19}{2}\ ft \times 10\ ft =$$

95 sq ft

—By a denominate number expressing square measurement

EXAMPLE: 3 ft × 6 sq ft

1.   Multiply the denominate numbers.
     (This includes the units of measure,
     such as ft × ft = sq ft and sq ft
     × ft = cu ft.)

3 ft × 6 sq ft = 18 cu ft

—By a denominate number expressing rate

EXAMPLE: 50 miles per hour × 3 hours

1.   Express the rate as a fraction using
     the fraction bar for *per.*

$$\frac{50\ miles}{1\ hour} \times \frac{3\ hours}{1} =$$

2.   Divide the numerator and denominator
     by any common factors, including
     units of measure.

$$\frac{50\ miles}{\underset{1}{\cancel{1\ hour}}} \times \frac{\overset{3}{\cancel{3\ hours}}}{1} =$$

3. Multiply numerators.
   Multiply denominators.

$$\frac{150 \text{ miles}}{1} =$$

4. Express the answer in the remaining unit.

150 miles

D.   DIVISION

—By a constant

EXAMPLE: 8 gal 3 qt ÷ 5

1. Express all denominate numbers
   in the same unit.

8 gal 3 qt = 35 qt

2. Divide the denominate number
   by the constant.

35 qt ÷ 5 = 7 qt

3. Express the answer using the
   largest possible units.

7 qt = 1 gal 3 qt

—By a denominate number expressing linear measurement

EXAMPLE: 11 ft 4 in ÷ 8 in

1. Express all denominate numbers
   in the same unit.

11 ft 4 in = 136 in

2. Divide the denominate numbers
   by a common factor. (This includes
   the units of measure, such as inches
   ÷ inches = 1.)

$$136 \text{ in} \div 8 \text{ in} =$$
$$\frac{\overset{17}{\cancel{136 \text{ in}}}}{\underset{1}{\cancel{8 \text{ in}}}} = \frac{17}{1} = 17$$

—By a linear measure with a square measurement as the dividend

EXAMPLE: 20 sq ft ÷ 4 ft

1. Divide the denominate numbers.

   (This includes the units of measure,
   such as sq ft ÷ ft = ft.)

$$20 \text{ sq ft} \div 4 \text{ ft}$$
$$\frac{\overset{5 \text{ ft}}{\cancel{20 \text{ sq ft}}}}{\cancel{4 \text{ ft}}} = \frac{5 \text{ ft}}{1}$$

2. Express the answer in the remaining unit.

5 ft

—By denominate numbers used to find rate

EXAMPLE: 200 mi ÷ 10 gal

1. Divide the denominate numbers.

$$\frac{\overset{20 \text{ mi}}{\cancel{200 \text{ mi}}}}{\underset{1 \text{ gal}}{\cancel{10 \text{ gal}}}} = \frac{20 \text{ mi}}{1 \text{ gal}}$$

2. Express the units with the fraction bar meaning *per.*

$$\frac{20 \text{ mi}}{1 \text{gal}} = \quad 20 \text{ miles per gallon}$$

Note: Alternate methods of performing the basic operations will produce the same results. The choice of method is determined by the individual.

SECTION II

UNITS AND TABLES OF ELECTRONIC MEASUREMENT

The field of electronics is made up of many complex units and numbers. In order to communicate correctly, it is important that each person in the field use the same language. A set of standard units and symbols is used. Table I lists the units and symbols most commonly used in electronics. These are used extensively in this workbook.

TABLE I QUANTITIES AND UNITS OF MEASUREMENT		
QUANTITY	UNIT	EXAMPLE
Capacitance (C)	farad (F)	C_1 = 0.000 01 F
Capacitive reactance (X_c)	ohm (Ω)	X_c = 45 Ω
Current (I)	ampere (A)	I_1 = 0.5 A
Frequency (f)	hertz (Hz)	f = 60 Hz
Impedance (Z)	ohm (Ω)	Z = 50 Ω
Inductance (L)	henry (H)	L_2 = 0.15 H
Inductive reactance (X_L)	ohm (Ω)	X_L = 37 Ω
Power (P)	watt (W)	P = 100 W
Reactance (X)	ohm (Ω)	X = 12 Ω
Resistance (R)	ohm (Ω)	R_2 = 15 Ω
Time (t)	second (s)	t = 0.8 s
Voltage (E)	volt (V)	E_s = 12 V

There is a need to make numbers like 68,000,000 ohms or 0.000 052 ampere easier to write. To simplify standard units, only specific powers of ten are used. The powers used most are 10^6, 10^3, 10^{-3}, 10^{-6}, and 10^{-12}.

MEGA: When mega is used as a prefix, it stands for one million or 10^6. For example, 5 megavolts means 5×10^6 volts or 5,000,000 volts. The value 15 megohms means 15×10^6 ohms or 15,000,000 ohms. The letter that indicates mega is M. Therefore, 5 megavolts is written 5 MV and 15 megohms is written 15 MΩ.

KILO: When kilo is used as a prefix, it stands for one thousand or 10^3 or 1,000. For example, 5.6 kilohms means 5.6×10^3 ohms or 5,600 ohms. The value 0.85 kilohertz means 0.85×10^3 hertz or 850 hertz. The letter used to indicate kilo is k. Therefore, 5.6 kilohms is 5.6 kΩ and 0.85 kilohertz is 0.85 kHz.

MILLI: When milli is used as a prefix, it stands for one thousandth or 0.001 or 10^{-3}. For example, 26 milliamperes means 26×10^{-3} ampere or 0.026 ampere. The value 8 milliwatts means 8×10^{-3} watt or 0.008 watt. The letter that indicates milli is m. Therefore, 26 milliamperes is 26 mA and 8 milliwatts is 8 mW.

MICRO: When micro is used as a prefix, it stands for 10^{-6}, 0.000 001, or one millionth. For example, 75 microhenries means 75×10^{-6} henry or 0.000 075 henry. The value 359 microseconds means 359×10^{-6} second or 0.000 359 second. The letter that indicates micro is μ. This is the Greek letter mu. Therefore, 75 microhenries is 75 μH and 359 microseconds is 359 μs.

PICO: When pico is used as a prefix, it stands for 10^{-12}. For example, 150 picofarads means 150×10^{-12} farad or 0.000 000 000 15 farad. The letter that indicates pico is p. Therefore, 150 picofarads is 150 pF.

TABLE II STANDARD ENGINEERING UNITS			
NUMBER	**POWER**	**PREFIX**	**LETTER**
1 000 000 000 000	10^{12}	terra	T
1 000 000 000	10^{9}	giga	G
1 000 000	10^{6}	mega	M
1 000	10^{3}	kilo	k
1	10^{0}		
0.001	10^{-3}	milli	m
0.000 001	10^{-6}	micro	μ
0.000 000 001	10^{-9}	nano	n
0.000 000 000 001	10^{-12}	pico	p

There are many times when units of measurement must be added, subtracted, multiplied, and divided. Operations such as these become easier with practice.

ADDITION AND SUBTRACTION: When units are added or subtracted, they must be the same units. To add 470 ohms and 1.2 kilohms, one of the values must be expressed as the other unit. This would mean 470 ohms plus 1,200 ohms or 0.47 kilohm plus 1.2 kilohms. The answer would be 1,670 ohms or 1.67 kilohms.

MULTIPLICATION AND DIVISION: There are many situations in which electronic units are multiplied and/or divided. Tables III and IV have been provided as guides in working Ohm's law and Power law problems. While these tables do not contain all possible combinations, they will assist the student in working basic problems. The student will want to refer to an electronics text for additional information on combination Ohm's law and Power law problems. Section 5 of this workbook provides numerous practice problems.

ACCEPTABLE UNITS: There are many units in which a given answer or value can be expressed. Assume you have a voltage of 0.25 volt. This can also be written as 250 millivolts, 250,000 microvolts, or 0.000 25 kilovolt. As a general rule, if there are more than three numbers preceding or following a decimal point, a change of units should be made. In this example, 0.25 volt and 250 millivolts are the most acceptable. Look at these other examples:

$$4,700 \text{ V} \quad = \quad 4.7 \text{ kV}$$

$$0.182\ 6 \text{ A} \quad = \quad 182.6 \text{ mA}$$

$$1,600,000\ \Omega \quad = \quad 1.6 \text{ M}\Omega$$

TABLE III OHM'S LAW UNITS		
OPERATION	**SYMBOL**	**RESULT**
amperes × ohms	A × Ω	volts (V)
amperes × kilohms	A × kΩ	kilovolts (kV)
amperes × megohms	A × MΩ	megavolts (MV)
milliamperes × ohms	mA × Ω	millivolts (mV)
milliamperes × kilohms	mA × kΩ	volts (V)
milliamperes × megohms	mA × MΩ	kilovolts (kV)
microamperes × ohms	μA × Ω	microvolts (μV)
microamperes × kilohms	μA × kΩ	millivolts (mV)
microamperes × megohms	μA × MΩ	volts (V)
volts ÷ ohms	V ÷ Ω	amperes (A)
volts ÷ kilohms	V ÷ kΩ	milliamperes (mA)
volts ÷ megohms	V ÷ MΩ	microamperes (μA)
millivolts ÷ ohms	mV ÷ Ω	milliamperes (mA)
millivolts ÷ kilohms	mV ÷ kΩ	microamperes (μA)
millivolts ÷ megohms	mV ÷ MΩ	nanoamperes (nA)
volts ÷ amperes	V ÷ A	ohms (Ω)
volts ÷ milliamperes	V ÷ mA	kilohms (kΩ)
volts ÷ microamperes	V ÷ μA	megohms (MΩ)
millivolts ÷ amperes	mV ÷ A	milliohms (mΩ)
millivolts ÷ milliamperes	mV ÷ mA	ohms (Ω)
millivolts ÷ microamperes	mV ÷ μA	kilohms (kΩ)

TABLE IV POWER LAW UNITS

OPERATION	SYMBOL	RESULT
amperes $\times$ volts	A $\times$ V	watts (W)
amperes $\times$ millivolts	A $\times$ mV	milliwatts (mW)
amperes $\times$ microvolts	A $\times$ μV	microwatts (μW)
milliamperes $\times$ volts	mA $\times$ V	milliwatts (mW)
milliamperes $\times$ millivolts	mA $\times$ mV	microwatts (μW)
milliamperes $\times$ microvolts	mA $\times$ μV	nanowatts (nW)
watts $\div$ volts	W $\div$ V	amperes (A)
watts $\div$ kilovolts	W $\div$ kV	milliamperes (mA)
watts $\div$ megavolts	W $\div$ MV	microamperes (μA)
milliwatts $\div$ volts	mW $\div$ V	milliamperes (mA)
milliwatts $\div$ millivolts	mW $\div$ mV	amperes (A)
watts $\div$ amperes	W $\div$ A	volts (V)
watts $\div$ milliamperes	W $\div$ mA	kilovolts (kV)
watts $\div$ microamperes	W $\div$ μA	megavolts (MV)
milliwatts $\div$ amperes	mW $\div$ A	millivolts (mV)
milliwatts $\div$ milliamperes	mW $\div$ mA	volts (V)
milliwatts $\div$ microamperes	mW $\div$ μA	kilovolts (kV)

TABLE V COMBINATION OHM'S LAW AND POWER LAW EQUATIONS

EQUATION	OPERATION		RESULT
$E = \sqrt{PR}$	voltage	$= \sqrt{\text{watts} \times \text{ohms}}$	volts (V)
$E = \dfrac{P}{I}$	voltage	$= \dfrac{\text{watts}}{\text{amperes}}$	volts (V)
$E = IR$	voltage	$=$ amperes $\times$ ohms	volts (V)
$I = \dfrac{E}{R}$	current	$= \dfrac{\text{volts}}{\text{ohms}}$	amperes (A)
$I = \dfrac{P}{E}$	current	$= \dfrac{\text{watts}}{\text{volts}}$	amperes (A)
$I = \sqrt{\dfrac{P}{R}}$	current	$= \sqrt{\dfrac{\text{watts}}{\text{ohms}}}$	amperes (A)
$P = IE$	power	$=$ amperes $\times$ volts	watts (W)
$P = I^2 R$	power	$=$ amperes^2 $\times$ ohms	watts (W)
$P = \dfrac{E^2}{R}$	power	$= \dfrac{\text{volts}^2}{\text{ohms}}$	watts (W)
$R = \dfrac{E}{I}$	resistance	$= \dfrac{\text{volts}}{\text{amperes}}$	ohms (Ω)
$R = \dfrac{E^2}{P}$	resistance	$= \dfrac{\text{volts}^2}{\text{watts}}$	ohms (Ω)
$R = \dfrac{P}{I^2}$	resistance	$= \dfrac{\text{watts}}{\text{amperes}^2}$	ohms (Ω)

Carbon and composition resistors are usually marked with colored bands. Each band stands for a specific digit, the number of zeros, or the resistor tolerance. The tolerance is the percent the resistor value varies from its indicated value.

	1st Band	2nd Band	3rd Band	4th Band
COLOR	1st Digit	2nd Digit	Number of Zeros	Tolerance
Black	–	0	0	–
Brown	1	1	1	1%
Red	2	2	2	2%
Orange	3	3	3	–
Yellow	4	4	4	–
Green	5	5	5	–
Blue	6	6	6	–
Violet	7	7	7	–
Gray	8	8	8	–
White	9	9	9	–
Gold	–	–	x.1	5%
Silver	–	–	x.01	10%
None	–	–	–	20%

TABLE VI RESISTOR COLOR CODE

SECTION III

ELECTRONIC SYMBOLS AND FORMULAS

COMPONENT	SYMBOL	COMPONENT	SYMBOL
Ammeter	—(A)—	Fuse	
Battery	–‖‖+	Ground	
Capacitor	—)‖—	Inductor	—ɷɷɷ—
Cell	– ‖ +	Lamp	—(⍵)—
Circuit Breaker		Light Emitting Diode (LED)	
Diode	—▶‖—	Ohmmeter	—(Ω)—
Electrolytic Capacitor	–)‖ +	Photoconductive Cell	
Field Effect Transistor		Photovoltaic Cell	– +

COMPONENT	SYMBOL	COMPONENT	SYMBOL
Potentiometer		Switch (PBNO) Push Button Normally Open	
Relay		Switch-Rotary	
Resistor		Thermistor	
Rheostat		Transformer	
Silicon Controlled Rectifier (SCR)		Transistor (NPN)	
Speaker		Transistor (PNP)	
Switch (SPST) Single Pole Single Throw		Voltmeter	
Switch (SPDT) Single Pole Double Throw		Wires Connected	
Switch (DPST) Double Pole Single Throw		Wires Not Connected	
Switch (DPDT) Double Pole Double Throw		Zener Diode	
Switch (PBNC) Push Button Normally Closed			

PERIMETERS

Rectangle
$$P = 2l + 2w$$ where P = perimeter
                         l = length
                         w = width

Circle
$$C = \pi d$$ where C = circumference
                       π = 3.1416
                       d = diameter

AREAS

Rectangle
$$A = lw$$ where A = area
                     l = length
                     w = width

Circle
$$A = \pi r^2$$ where A = area
                         π = 3.1416
                         r = radius

Triangle
$$A = \tfrac{1}{2}bh$$ where A = area
                                 b = base
                                 h = height

Ellipse
$$A = \pi ab$$ where A = area
                         a = semimajor axes
                         b = semiminor axes
                         π = 3.1416

VOLUMES

Rectangular Solid
$$V = lwh$$ where V = volume
                      l = length
                      w = width
                      h = height

Cylinder
$$V = \pi r^2 l$$ where V = volume
                           π = 3.1416
                           r = radius
                           l = length

OHM'S LAW AND POWER LAW

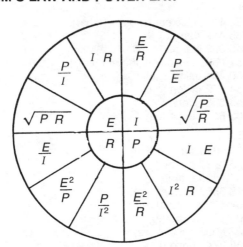

where P = power in watts
        I = current in amperes
        R = resistance in ohms
        E = voltage in volts

SECTION IV

MATHEMATICAL REFERENCE TABLES
ENGLISH-METRIC EQUIVALENTS

LENGTH MEASURE

1 inch (in)	≈	25.4 millimeters (mm)
1 inch (in)	≈	2.54 centimeters (cm)
1 foot (ft)	≈	0.304 8 meter (m)
1 yard (yd)	≈	0.914 4 meter (m)
1 mile (mi)	≈	1.609 kilometers (km)
1 millimeter (mm)	≈	0.039 37 inch (in)
1 centimeter (cm)	≈	0.393 70 inch (in)
1 meter (m)	≈	3.280 84 feet (ft)
1 meter (m)	≈	1.093 61 yards (yd)
1 kilometer (km)	≈	0.621 37 mile (mi)

AREA MEASURE

1 square inch (sq in)	≈	645.16 square millimeters (mm^2)
1 square inch (sq in)	≈	6.451 6 square centimeters (cm^2)
1 square foot (sq ft)	≈	0.092 903 square meter (m^2)
1 square yard (sq yd)	≈	0.836 127 square meter (m^2)
1 square millimeter (mm^2)	≈	0.001 550 square inch (sq in)
1 square centimeter (cm^2)	≈	0.155 00 square inch (sq in)
1 square meter (m^2)	≈	10.763 910 square feet (sq ft)
1 square meter (m^2)	≈	1.195 99 square yards (sq yd)

VOLUME MEASURE FOR SOLIDS

1 cubic inch (cu in)	≈	16.387 064 cubic centimeters (cm^3)
1 cubic foot (cu ft)	≈	0.028 317 cubic meter (m^3)
1 cubic yard (cu yd)	≈	0.764 555 cubic meter (m^3)
1 cubic centimeter (cm^3)	≈	0.061 024 cubic inch (cu in)
1 cubic meter (m^3)	≈	35.314 667 cubic feet (cu ft)
1 cubic meter (m^3)	≈	1.307 951 cubic yards (cu yd)

VOLUME MEASURE FOR FLUIDS

1 gallon (gal)	≈	3,785.411 cubic centimeters (cm^3)
1 gallon (gal)	≈	3.785 411 liters (L)
1 quart (qt)	≈	0.946 353 liter (L)
1 ounce (oz)	≈	29.573 530 cubic centimeters (cm^3)
1 cubic centimeter (cm^3)	≈	0.000 264 gallon (gal)
1 liter (L)	≈	0.264 172 gallon (gal)
1 liter (L)	≈	1.056 688 quarts (qt)
1 cubic centimeter (cm^3)	≈	0.033 814 ounce (oz)

MASS MEASURE

1 pound (lb)	≈	0.453,592 kilogram (kg)
1 pound (lb)	≈	453.592 37 grams (g)
1 ounce (oz)	≈	28.349 523 grams (g)
1 ounce (oz)	≈	0.028 350 kilogram (kg)
1 kilogram (kg)	≈	2.204 623 pounds (lb)
1 gram (g)	≈	0.002 205 pound (lb)
1 kilogram (kg)	≈	35.273 962 ounces (oz)
1 gram (g)	≈	0.035 274 ounce (oz)

METRIC RELATIONSHIPS

The base units in SI metrics include the meter and the gram. Other units of measure are related to these units. The relationship between the units is based on powers of ten and uses these prefixes:

kilo (1,000)   hecto (100)   deka (10)   deci (0.1)   centi (0.01)   milli (0.001)

These tables show the most frequently used units with an asterisk (*).

METRIC LENGTH

10 millimeters (mm)*	=	1 centimeter (cm)*
10 centimeters (cm)	=	1 decimeter (dm)
10 decimeters (dm)	=	1 meter (m)*
10 meters (m)	=	1 dekameter (dam)
10 dekameters (dam)	=	1 hectometer (hm)
10 hectometers (hm)	=	1 kilometer (km)*

- ◆ To express a metric length unit as a smaller metric length unit, multiply by a positive power of ten such as 10, 100, 1,000, 10,000 etc.
- ◆ To express a metric length unit as a larger metric length unit, multiply by a negative power of ten such as 0.1, 0.01, 0.001, 0.000 1, etc.

METRIC AREA MEASURE

100 square millimeters (mm^2)	=	1 square centimeter (cm^2)*
100 square centimeters (cm^2)	=	1 square decimeter (dm^2)
100 square decimeters (cm^2)	=	1 square meter (m^2)*
100 square meters (m^2)	=	1 square dekameter (dam^2)
100 square dekameters (dam^2)	=	1 square hectometer (hm^2)*
100 square hectometers (hm^2)	=	1 square kilometer (km^2)

- ◆ To express a metric area unit as a smaller metric area unit, multiply by 100, 10,000, 1,000,000, etc.
- ◆ To express a metric area unit as a larger metric area unit, multiply by 0.01, 0.000 1, 0.000 001, etc.

METRIC VOLUME MEASURE FOR SOLIDS

1 000 cubic millimeters (mm^3)	=	1 cubic centimeter (cm^3)*
1 000 cubic centimeters (cm^3)	=	1 cubic decimeter (dm^3)*
1 000 cubic decimeters (dm^3)	=	1 cubic meter (m^3)*
1 000 cubic meters (m^3)	=	1 cubic dekameter (dam^3)
1 000 cubic dekameters (dam^3)	=	1 cubic hectometer (hm^3)
1 000 cubic hectometers (hm^3)	=	1 cubic kilometer (km^3)

- ◆ To express a metric volume unit for solids as a smaller metric volume unit for solids, multiply by 1,000, 1,000,000, 1,000,000,000, etc.
- ◆ To express a metric volume unit for solids as a larger metric volume unit for solids, multiply by 0.001, 0.000 001, 0.000 000 001, etc.

METRIC VOLUME MEASURE FOR FLUIDS

10 milliliters (mL)*	=	1 centiliter (cL)
10 centiliters (cL)	=	1 deciliter (dL)
10 deciliters (dL)	=	1 liter (L)*
10 liters (L)	=	1 dekaliter (daL)
10 dekaliters (daL)	=	1 hectoliter (hL)
10 hectoliters (hL)	=	1 kiloliter (kL)

- ◆ To express a metric volume unit for fluids as a smaller metric volume unit for fluids, multiply by 10, 100, 1,000, 10,000, etc.
- ◆ To express a metric volume unit for fluids as a larger metric volume unit for fluids, multiply by 0.1, 0.01, 0.001, 0.000 1, etc.

METRIC VOLUME MEASURE EQUIVALENTS

1 cubic decimeter (dm^3)	=	1 liter (L)
1 000 cubic centimeters (cm^3)	=	1 liter (L)
1 cubic centimeter (cm^3)	=	1 milliliter (mL)

METRIC MASS MEASURE

10 milligrams (mg)*	=	1 centigram (cg)
10 centigrams (cg)	=	1 decigram (dg)
10 decigrams (cg)	=	1 gram (g)*
10 dekagrams (dag)	=	1 hectogram (hg)
10 hectograms (hg)	=	1 kilogram (kg)*
1 000 kilograms (kg)	=	1 megagram (Mg)*

◆ To express a metric mass unit as a smaller metric mass unit, multiply by 10, 100, 1,000, 10,000, etc.

◆ To express a metric mass unit as a larger metric mass unit, multiply by 0.1, 0.01, 0.001, 0.000 1, etc.

Metric measurements are expressed in decimal parts of a whole number. For example, one half millimeter is written as 0.5 mm.

In calculating with the metric system, all measurements are expressed using the same prefixes. If answers are needed in millimeters, all parts of the problem should be expressed in millimeters before the final solution is attempted. Diagrams that have dimensions in different prefixes must first be expressed using the same unit.

METRIC AND CUSTOMARY DECIMAL EQUIVALENTS
FOR FRACTIONAL PARTS OF AN INCH

Fraction	Decimal Equivalent Customary (in)	Metric (mm)	Fraction	Decimal Equivalent Customary (in)	Metric (mm)
1/64—0.015 625		0.396 9	33/64—0.515 625		13.096 9
1/32———0.031 25		0.793 8	17/32———0.531 25		13.493 8
3/64—0.046 875		1.190 6	35/64—0.546 875		13.890 6
1/16———0.062 5		1.587 5	9/16———0.562 5		14.287 5
5/64—0.078 125		1.984 4	37/64—0.578 125		14.684 4
3/32———0.093 75		2.381 3	19/32———0.593 75		15.081 3
7/64—0.109 375		2.778 1	39/64—0.609 375		15.478 1
1/8———0.125 0		3.175 0	5/8———0.625 0		15.875 0
9/64—0.140 625		3.571 9	41/64-0.640 625		16.271 9
5/32———0.156 25		3.968 8	21/32———0.656 25		16.668 8
11/64—0.171 875		4.365 6	43/64-0.671 875		17.065 6
3/16———0.187 5		4.762 5	11/16———0.687 5		17.462 5
13/64-0.203 125		5.159 4	45/64-0.703 125		17.859 4
7/32———0.218 75		5.556 3	23/32———0.718 75		18.256 3
15/64-0.234 375		5.953 1	47/64-0.734 375		18.653 1
1/4———0.250		6.350 0	3/4———0.750		19.050 0
17/64-0.265 625		6.746 9	49/64-0.765 625		19.446 9
9/32———0.281 25		7.143 8	25/32———0.781 25		19.843 8
19/64-0.296 875		7.540 6	51/64-0.796 875		20.240 6
5/16———0.312 5		7.937 5	13/16———0.812 5		20.637 5
21/64-0.328 125		8.338 4	53/64-0.828 125		21.034 4
11/32———0.343 75		8.731 3	27/32———0.843 75		21.431 3
23/64-0.359 375		9.128 1	55/64-0.859 375		21.828 1
3/8———0.375 0		9.525 0	7/8———0.875 0		22.225 0
25/64-0.390 625		9.921 9	57/64-0.890 625		22.621 9
13/32———0.406 25		10.318 8	29/32———0.906 25		23.018 8
27/64-0.421 875		10.715 6	59/64-0.921 875		23.415 6
7/16———0.437 5		11.112 5	15/16———0.937 5		23.812 5
29/64-0.453 125		11.509 4	61/64-0.953 125		24.209 4
15/32———0.468 75		11.906 3	31/32———0.968 75		24.606 3
31/64-0.484 375		12.303 1	63/64-0.984 375		25.003 1
1/2———0.500		12.700 0	1———1.000		25.400 0

Glossary

Alternating current (AC) — An electric current that reverses direction at regularly recurring intervals.

Ammeter — An instrument for measuring electric current.

Ampere (A) — A unit of electric current.

Amplifier — A device used to increase the strength of an input current or voltage signal so that the output is stronger, but has the same features as the input.

AND gate — A binary circuit with two or more inputs and a single output. The output logic is 1 when input logic is 1.

Bandwidth — A group or band of frequencies that surround a center frequency.

Base current (I_b) — The current through the base of a transistor.

Base voltage (E_b) — The voltage present at the base of a transistor.

Battery — Two or more cells in one case producing a pure DC voltage.

Beta (β) — *See* Current gain.

Binary number system — A numbering system using the number 2 as a base.

Boolean algebra — An algebraic system dealing with on/off circuit elements. The system is named for George Boole who introduced it in 1847.

Capacitance (C) — The property of a capacitor that permits the storage of electrostatic energy. The unit of capacitance is the farad.

Capacitive reactance (X_c) — The opposition to an alternating current due to capacitance. The capacitive reactance is expressed in ohms.

Capacitor — An electrical device consisting of two conducting surfaces that are oppositely charged and are separated by thin layers of dielectric material.

Circuit — The conducting part, or a system of conducting parts, through which an electrical current passes.

Coil — *See* Inductor.

Collector current (I_c) — The current through the collector of a transistor.

Collector voltage (E_c) — The voltage present at the collector of a transistor.

Component — An individual part or device such as a resistor, a transistor, or an integrated circuit.

Conductance (G) — The ability of material to carry electrical current. It is measured in siemens.

Current (I) — The transfer of an electric charge through a material. Current is measured in amperes.

Current gain (β) — The ratio of output current to input current of a common emitter transistor.

Cycle — A complete positive and a complete negative alternation of voltage or current.

Cycles per second — *See* Hertz.

Digital — A device or circuit in which the output varies in discrete steps (on/off).

Diode — A two-terminal solid state device that conducts current in one direction but not in the other.

Direct current (DC) — An electrical current in which there is a continuous transfer of charge in one direction only.

Emitter current (I_e) — The current through the emitter of a transistor.

Emitter voltage (E_e) — The voltage present at the emitter of a transistor.

Etch — Removing the conductive material on a printed circuit board using chemical action.

Farad (F) — The unit of capacitance.

Field effect transistor (FET) — A type of transistor with a very high input impedance.

Filter — A circuit or device that blocks some frequencies and passes others.

Frequency (f) — The number of complete cycles in a unit of time. It is expressed in hertz.

Fuse — An electrical safety device that is constructed of wire or fusible material. It melts and opens an electrical circuit when the rated amount of current is exceeded.

Gate — A circuit with one or more inputs and one output that is dependent upon the signal or combination of signals at the input(s).

Henry (H) — The unit of inductance.

Hertz (Hz) — The unit for frequency; cycles per second.

Impedance (*Z*) — The total opposition to an alternating current. Included are inductive reactance, capacitive reactance, and resistance. It is expressed in ohms.

Inductance (*L*) — The property of an alternating current circuit to induce an electromotive force by varying the current. It is expressed in henries.

Inductive reactance (*X_L*) — The opposition of an inductor to an alternating current. It is expressed in ohms.

Inductor — An electrical device that acts upon another or is itself acted upon by induction. As the conductor is wound into a spiral or coil, the inductive intensity increases.

Integrated circuit (IC) — A combination of devices and elements in a miniature self-contained package. Usually ICs are designed to perform a particular function.

Kirchoff's current law — The sum of the currents entering a junction equals the sum of the currents leaving that junction.

Kirchoff's voltage law — The sum of the voltage drops around a closed loop equals the applied voltage.

Load — The resistance connected across the output of a circuit.

Logic — A "logic" input refers to an input (1 or 0) to a gate in a digital circuit.

Mutual inductance (*M* or *L_m*) — The condition that exists in a circuit when the positions of the two inductors cause magnetic lines of force from one inductor to link the turns of the other.

NAND gate — A binary circuit with two or more inputs and a single output. The NAND gate consists of an AND gate followed by a NOT gate.

NOR gate — A binary circuit with two or more inputs and a single output. The NOR gate consists of an OR gate followed by a NOT gate.

NOT gate — A binary circuit with a single input and a single output. The output is a logic 1 when the input is a logic 0.

Ohm (Ω) — A unit of resistance.

Ohm's law — Current is directly proportional to the voltage and inversely proportional to the resistance.

Ohmmeter — A meter used to measure resistance in ohms.

Operational amplifier (op-amp) — An amplifier that can be used for many applications. Some of these are mathematical, such as addition, subtraction, and multiplication.

OR gate — A binary circuit with two or more inputs and a single output. The output is a logic 1 when any input equals a logic 1.

Oscillator — A circuit for producing alternating current power at a frequency determined by the values of certain constants in the circuit.

Oscillosope — A test instrument using a cathode ray tube (CRT) to display signal patterns.

Parallel circuit — A method of connecting a circuit so the current has two or more paths to follow.

Peak — The point of maximum voltage, current, or power during a cycle.

Peak-to-peak — The voltage, current, or power measured from positive peak to negative peak.

Period — The length of time for one cycle to occur.

Photoconductive cell — A resistor with a higher resistance in the dark and a lower resistance when exposed to light.

Photovoltaic cell — A cell that generates a voltage when struck by light.

Potentiometer — A three-terminal variable resistor.

Power (*P*) — The rate of doing work. It is expressed in watts.

Power gain — The ratio of output power to input power.

Reactance (*X*) — Opposition to alternating current due to inductance and/or capacitance. It is expressed in ohms.

Relay — An electromechanical switch.

Resistance (*R*) — The opposition a material has to the flow of charge. It is expressed in ohms.

Resistor — A device that opposes the flow of an electric current. It is used for protection, operation, or current control.

Rheostat — A two-terminal variable resistor.

rms voltage (E_{rms}) — The DC equivalent of an AC sine wave.

Schematic — The diagram of an electrical or electronic circuit.

Series circuit — A method of connecting a circuit so the current has one path to follow.

Solar cell — *See* Photovoltaic cell.

Solder — An alloy of lead and tin that melts at a low temperature. It is used to join metallic surfaces.

Thermistor — A resistor that changes resistance when exposed to temperature change.

Tolerance — The amount of deviation from a specific value that is allowed.

Transformer — An electromagnetic device using induction to increase or decrease alternating current voltage.

Transistor — An electronic device consisting of a small block of semiconductors with a minimum of three electrodes.

Volt (V) — A unit of electrical potential or pressure.

Voltage (E) — The electromotive force or electrical pressure. It is expressed in volts.

Voltage gain — The ratio of output voltage to input voltage.

Voltmeter — A meter used to measure voltage.

Watt (W) — A unit of power.

Wattmeter — A meter used to measure power.

Zener diode — A diode that maintains a constant breakdown voltage.

ANSWERS TO ODD-NUMBERED PRACTICAL PROBLEMS

UNIT 1 ADDITION OF WHOLE NUMBERS

1. 570 parts
3. 223 components
5. 180 m
7. 56 V

9. 5,021 km
11. 1,600,000 Hz
13. 1,298 pF

15. 213 mH
17. 55,497 speakers
19. 28,100 Ω

UNIT 2 SUBTRACTION OF WHOLE NUMBERS

1. 1,163 bulbs
3. 577 transistors
5. 6,500 Ω
7. 71,463 boards

9. 13 m
11. $133
13. 63,671 ft

15. $675
17. 15V
19. 113 V

UNIT 3 MULTIPLICATION OF WHOLE NUMBERS

1. 84 in
3. 4,875 µA
5. a. $90
   b. $270

7. 24 V
9. 6,960 ft
11. 125 W
13. 100 mV

15. 192,500 cycles
17. 500 mV
19. 2,688 songs

UNIT 4 DIVISION OF WHOLE NUMBERS

1. 52 mph
3. a. $96
   b. $12
5. a. 148 sets
   b. $296

5. c. 2
7. 552 mi
9. a. 720 d
   b. 144 w
11. 14 d

13. 13 gal
15. 150 mA
17. 100 V
19. 6 A

UNIT 5 COMBINED OPERATIONS WITH WHOLE NUMBERS

1. 80,965 circuits
3. $2,771
5. 15 Ω
7. 12 V

9. 22 V
11. 35 mi
13. 125,000 Ω
15. 10,500 Ω

17. 240 in
19. 5,475
21. 89,280,000 mi

UNIT 6 ADDITION OF COMMON FRACTIONS

1. $11\frac{1}{4}$ h
3. $42\frac{3}{4}$ h
5. $107\frac{11}{24}$ V
7. $6\frac{1}{6}$ gal
9. $2\frac{3}{4}$ in
11. $41\frac{19}{32}$ in
13. 10 in
15. $\frac{19}{64}$ in
17. $65\frac{3}{4}$ in
19. $\frac{5}{16}$ in

UNIT 7 SUBTRACTION OF COMMON FRACTIONS

1. $30\frac{3}{4}$ ft
3. $\frac{5}{8}$ A
5. $\frac{5}{16}$ in
7. $\frac{3}{16}$ in
9. a. $\frac{13}{16}$ in
   b. $\frac{7}{16}$ in
   c. $\frac{1}{2}$ in
   d. $\frac{3}{16}$ in
   e. $\frac{3}{16}$ in
   f. $\frac{3}{8}$ in
11. $1\frac{1}{8}$ lb
13. 8 mi
15. $\frac{11}{64}$ in
17. $1\frac{1}{2}$ V
19. $3\frac{1}{4}$ A

UNIT 8 MULTIPLICATION OF COMMON FRACTIONS

1. $41\frac{1}{4}$ h
3. $3\frac{1}{2}$ in
5. $110\frac{5}{8}$ in
7. $5\frac{9}{10}$ W
9. 68 V
11. $170\frac{1}{2}$ mi
13. $105\frac{3}{4}$ in
15. $10\frac{4}{5}$ V
17. a. $131\frac{1}{4}$ h
    b. $376\frac{1}{4}$ h
19. 3,486 in

UNIT 9 DIVISION OF COMMON FRACTIONS

1. 320 wires
3. 51 mph
5. $\frac{27}{32}$ in
7. $4/lb
9. $3\frac{7}{8}$ A
11. $21\frac{1}{2}$ Ω
13. 45 r/min
15. 18 mi/gal
17. 32
19. $22/h

UNIT 10 COMBINED OPERATIONS WITH COMMON FRACTIONS

1. 14 mi/gal
3. a. $77\frac{3}{4}$ in
   b. $\frac{1}{4}$ in
5. 14 Ω
7. $\frac{1}{5}$ A
9. $\frac{3}{20}$ A
11. $410.67
13. $231
15. 155
17. $3\frac{5}{8}$ pt
19. 605 ft
21. 88 V

UNIT 11 ROUNDING NUMBERS

1. 12.88 V
3. 485.6 Ω
5. 1,414.837 Hz
7. 0.240 65 A
9. 0.9 A
11. 2,800 Ω
13. 0.4 mA
15. 153,730 Hz
17. 0.087 A
19. 0.000 004 A

UNIT 12 CONVERSION OF COMMON AND DECIMAL FRACTIONS

1. 12¼
3. 19¹⁄₅₀
5. 5⅞
7. 0.875 in

9. 7.093 75 in
11. ⁵¹⁄₅₀₀ in
13. 0.125 A

15. ½
17. 0.875
19. ⁷⁄₂₀ V

UNIT 13 ADDITION OF DECIMAL FRACTIONS

1. 1.23 in
3. 72.97 V
5. 0.25 in
7. 0.940 5 in

9. 1.97 in
11. 6.81 L
13. 95.58 mi

15. $298.86
17. 0.032 3 in
19. 1,161.14 Ω

UNIT 14 SUBTRACTION OF DECIMAL FRACTIONS

1. 4.719 in
3. 1.53 ft
5. 3.812 5 in
7. 0.75 in

9. 1.062 5 in
11. 7.298 V
13. 0.527 25 A

15. 6.481 V
17. 1.24 cm
19. $65.05

UNIT 15 MULTIPLICATION OF DECIMAL FRACTIONS

1. $1,338.75
3. 118.069 V
5. 169.68 V
7. 219.877 V

9. 132 mi
11. 1,575 cycles
13. 0.75 Ω

15. a. 36 V
     b. 0.18 W
17. 7,440 mi
19. 10 V

UNIT 16 DIVISION OF DECIMAL FRACTIONS

1. 42
3. 32 songs
5. 186,000 mi/s
7. 0.016 7 s

9. 2,000 Hz
11. $3.85
13. 7

15. 80 Ω
17. 25 V
19. 0.02 A

UNIT 17 COMBINED OPERATIONS WITH DECIMAL FRACTIONS

1. $544.35
3. 3.416 V
5. $227.61
7. 125 Hz
9. 16.968 V

11. a. 120 V
     b. 60 V
     c. 42.42 V
13. 8 Ω

15. 0.002 5 A
17. 0.4 A
19. 5.76 Ω
21. 100.85 ft

UNIT 18 EXPONENTS

1. 1,000,000	9. 130,321	15. 3,175.2 CM
3. 576	11. 20.25 sq in	17. 74,088 cm^3
5. 9.7	13. 225 CM	19. 2,645 sq in
7. 17$\frac{1}{64}$ or 17.0156		

UNIT 19 SCIENTIFIC NOTATION

1. 1×10^{17}	9. 5×10^9	15. $1.545\ 455 \times 10^4\ \Omega$
3. 6.325×10^{-3}	11. 2.5×10^1 V	17. $3.676\ 5 \times 10^{-4}$ A
5. 3×10^{-6}	13. $2.5 \times 10^4\ \Omega$	19. 2.5×10^{-4} A
7. 3.6×10^3		

UNIT 20 EQUIVALENT ELECTRONIC UNITS

1. 96,000,000 Hz	9. 300 mV	15. 50,000 μs
3. 470,000 Ω	11. 47.52 kHz	17. 5.6 kΩ
5. 2,000 pF	13. 8,000 μH	19. 9,400,000 kHz
7. 0.25 mA		

UNIT 21 OPERATIONS WITH ELECTRONIC UNITS

1. 11.45 kΩ	9. 3.35 mA	17. 5.45 MHz
3. 1.95 mA	11. 1,200 kHz	19. 8.1 V
5. 2.45 H	13. 6 V	21. 166.2 A
7. 1,060.5 mV	15. 0.282 8 V	

UNIT 22 ROOTS

1. 3	9. 6	15. 8.6
3. 7	11. 28 cm	17. 10 mils
5. 3	13. 12 V	19. 4$\frac{3}{4}$ in or 4.75 in
7. 4		

UNIT 23 EQUATIONS

1. $x = 4$	9. $a = 3.25$	17. $a = 42.5$
3. $x = 125$	11. $c = 1.16$	19. $x = 35$
5. $p = 13.8$	13. $b = 24$	21. $z = 0.5$
7. $y = 34.6$	15. $y - 31$	23. $z = 0$

25. $x = 7.92$

27. $R = E/I$

29. $R = 1/G$

31. $f = X_L/2\pi L$

33. $L_m = (L_t - L_1 - L_2)/2$

35. $R = T_{max}/5C$

37. $X_L = QR$

39. $f = 300,000,000/\lambda$

41. $E_2 = E_S - E_1 - E_3$

43. $°C = \frac{5}{9}(°F - 32)$

45. $1/R_1 = 1/R_t - 1/R_2$

47. $x = (y - b)/m$

49. $I_1 = (I_2 \times R_2)/R_1$

UNIT 24 FORMULAS

1. 2.507 Ω

3. 0.005 S

5. 5 ms

7. 0.025 F

9. 16

11. 120.19 V

13. 33.946 V

15. 30 mH

17. 800 Hz

19. 80

21. 279 mH

23. a.   1.2 A

    b.   1.8 A

25. 12 µF

UNIT 25 OHM'S LAW

1. 0.672 V or 672 mV

3. 0.0128 A or 12.8 mA

5. 0.0011 A or 1.1 mA

7. 10.88 V

9. 0.4 kΩ or 400 Ω

11. 0.035 A or 35 mA

13. 4.1 V

15. 2.703 kΩ (2,703 Ω)

17. 0.0706 mA or 70.6 µA

19. 2,500 mV or 2.5 V

21. 46.875 kΩ (46,875 Ω)

23. 3.75 mA

25. 20 V

UNIT 26 POWER LAW

1. 900 mW or 0.9 W

3. 4 V (0.004 kV)

5. 20 mA or 0.02 A

7. a.   No

    b.   1.968 W (1,968 mW)

9. 3.125 V

11. 0.417 A or 417 mA

13. 2.4 µW

15. 38.4 mW (38,400 µW)

UNIT 27 COMBINATION OHM'S LAW AND POWER LAW PROBLEMS

1. 18 Ω

3. 2 A

5. 0.2 kΩ or 200 Ω

7. 429 µA

9. 7.2 kΩ

11. 0.09 A or 90 mA

13. 54 kΩ

15. 0.5 A

17. 0.1 kΩ or 100 Ω

19. 37.5 mA (0.0375 A)

21. a.   0.24 MΩ or 240 kΩ

21. b.   600 µW

23. a.   8.4 V (8,400 mV)

    b.   1.26 mW (0.00126 W)

25. a.   8 Ω

    b.   2 V

UNIT 28 RATIO

1. 1:8

3. 5:2

5. 2:3

7. 2:9

9. 1:36

11. 8:1

13. 6,000:1
15. 2:17

17.   135:2

19.   50:1

UNIT 29 PROPORTION

1. 16
3. 0.15 A
5. 13.5
7. 22
9. 1
11. 0.1 V
13. 2.5 A

15.   3.6 h
17.   3.2 A
19.   2.78 Ω
21.   3,500 ft
23.   5.6 kΩ
25.   a.   8.64 in

25.   b.   8.81 in
    c.   9.23 in
    d.   9.73 in
    e.   4.78 in
    f.   3.86 in
    g.   3.86 in

UNIT 30 LINEAR MEASURE

1. 3.22 km
3. 20.34 ft
5. 1,520 mm
7. 3,937.00 ft
9. 80½ in

11.   324 in
13.   294¾ in
15.   48 ft 1½ in
17.   64.36 km

19.   120.65 mm
21.   14 in
23.   299,274 km
25.   328.1 ft

UNIT 31 SURFACE AND VOLUME MEASURE

1. 864 sq in
3. 2,626.38 sq mi
5. 1,575 cm^2
7. 64 cu in
9. 852 sq in

11.   3.14 in
13.   52,646.93 cu in
15.   113,097.6 cm^3
17.   4 cu ft
19.   8 in

21.   a.   14,400 W
    b.   2,251.8 lb
    c.   22,518 Btu
    d.   31,918 Btu
    e.   0.705 hr or 42.3 min

UNIT 32 GRAPHS

1. 2 mA
3. a.   2 kΩ
   b.   2 kΩ
   c.   2 kΩ
5. 0 mA
7. 15 Ω to 18 Ω

9.   16 mW
11.   10 mA
13.   1,000 Ω
15.   200
17.   30 V

19.   15 kHz and 85 kHz
21.   75 kHz
23.   30° and 150°
25.   −28 V

UNIT 33 PERCENTAGES AND AVERAGES

1. 7.5

3.   40.7

5.   0.788

7. 55%	13. 2.56%	19. 1.99 V
9. 23.93 V	15. 6	21. 8.7%
11. 39 h/wk	17. 5 W	

UNIT 34 TOLERANCES AND COLOR CODE

1. orange, orange, red, gold	11.	5,250 Ω	17.	a.	13.5 µF
3. brown, green, green, silver	13.	a. 96 mH		b.	27 µF
5. yellow, violet, orange, silver		b. 144 mH	19.	a.	824.5 kHz
7. 12 MΩ 10%	15.	a. 9.27 kΩ		b.	875.5 kHz
9. 2.4 MΩ 5%		b. 11.33 kΩ			

UNIT 35 RIGHT TRIANGLES

1. 10	7. 125	13. 10.61 in
3. 13.42	9. 45.69	15. 72 in
5. 8.93	11. 91.39 ft	

UNIT 36 TRIGONOMETRIC FUNCTIONS

1. 0.438 4	9. 86°	15. ∠A = 28° ∠B = 62°
3. 0.034 9	11. ∠A = 38° ∠B = 52°	17. ∠A = 42° ∠B = 48°
5. 2.144 5	13. ∠A = 45° ∠B = 45°	19. ∠A = 10° ∠B = 80°
7. 71°		

UNIT 37 PLANE VECTORS

1. 36.06 Ω	7. 250 V	13. 5 A
3. 17.32 Ω	9. 127.37 V	15. 119.64 V
5. 22.80 V	11. 120.2 V	

UNIT 38 INDUCTIVE CIRCUITS AND RL CIRCUITS

1. 50.27 kΩ	7. 19.2 kΩ	13. 0.2 mA
3. 5.00 kHz	9. 5,389 Ω	15. 6 V (6,000 mV)
5. 398 mH	11. 21 kΩ	

UNIT 39 CAPACITIVE CIRCUITS AND RC CIRCUITS

1. 31.8 Ω	5. 0.035 µF	9. a. 530.5 Ω
3. 6.37 kΩ	7. 62.5 Ω	b. 1,132.0 Ω

11. 16 kΩ

13. 0.1 A

15. a. 382.6 Ω

    b. 457.0 Ω

15. c. 1.2 V

UNIT 40 RLC CIRCUITS

1. 3.8 kΩ

3. 1,508.0 Ω

5. 181.7 Ω

7. 47.7 mH

9. 2.4 kΩ

11. 3.0 kΩ

13. 226.2 Ω

15. 39.1 Ω

17. 230.466 Ω

19. 11.314 mA

UNIT 41 SIMULTANEOUS EQUATIONS

1. $x = 21, y = 7$

3. $a = 16, b = -4$

5. $z = -15, j = 45$

7. $w = 18, z = 22$

9. $I_1 = 0.215$ A $I_2 = 0.0844$ A $I_3 = 0.299$ A

    $E_1 = 2.954$ V $E_2 = 3.87$ V $E_3 = 7.475$ V